VASCULAR PLANT FAMILIES

VASCULAR PLANT FAMILIES

AN INTRODUCTION TO THE FAMILIES OF VASCULAR
PLANTS NATIVE TO NORTH AMERICA AND SELECTED
FAMILIES OF ORNAMENTAL OR ECONOMIC IMPORTANCE

By

JAMES PAYNE SMITH, JR.

Illustrations By Kathryn E. Simpson

James Payne Smith, Jr.
Associate Professor of Botany
Director of the Herbarium
Humboldt State University
Arcata, California

MAD RIVER PRESS INC.

©1977 James P. Smith, Jr.
Published by Mad River Press, Inc.,
                    Route 1, Box 151 — B
                    Eureka, California 95501
Printed by Eureka Printing Company Inc.,
                    Eureka, California 95501
ISBN 0-916422-07-0

Dedicated to my Mother and Father

TABLE OF CONTENTS

Introduction

This book is intended for the student who is taking an elementary plant taxonomy course or for anyone who is interested in learning about the families of vascular plants. It is assumed that the reader is unfamiliar with the subject.

My primary objective in writing this text is to provide a useful treatment of the families of vascular plants of North America, with special emphasis on the flora of the United States. All of the families with native species are treated. In addition, I have quite arbitrarily included a few other families of ornamental or economic importance. The more commonly encountered families are accorded individual treatments. In writing these descriptions, I have attempted a compromise between a treatment which is so meagre as to be worthless and one which is so detailed that the reader is soon lost in trivial exceptions. Those desiring more encyclopedic presentations should consult Hutchinson (1967, 1969, 1973), Lawrence (1951), Melchior and Werdermann (1954), and Willis (1973). The features of more obscure families are summarized in comparison tables at the beginning of each subclass.

The sequence of divisions, classes, subclasses, and orders follows that of Cronquist, Takhtajan, and Zimmermann (1966), except for the ending of division names. The family sequence for the flowering plants is that of Cronquist (1968). I have departed from it in a few instances. This system has much to recommend it, and even at the introductory level seems highly preferable to the more widely used scheme of Engler or some artificial one.

The spelling of several family names differs slightly from that of older texts because of recent codification of these names. Many of the common names used for families do enjoy popular usage, while others are admittedly shorthand versions of semitechnical names which taxonomists employ when chatting with one another. The estimates of family size are taken from Melchior and Werdermann (1954) and Willis (1973). The number of genera and species in a family appears as a "fraction" at the beginning of the second paragraph of a family treatment. The figure 32/4500 should be read, 32 genera and 4500 species. The number in parentheses after a generic name is an estimate of the number of species on a world wide basis. I have attempted in the "Recognition Characters" to extract from the more complete descriptions those salient taxonomic features which might be helpful to the beginning student. The conventions used in writing the floral formulae used in these sections, along with those which appear in the comparisons of families at the beginning of each subclass are explained in Chapter 5.

I wish to thank the students in my plant taxonomy classes at Humboldt State University who first encouraged me to attempt a textbook on plant families and who pointed out errors and offered suggestions after the first edition appeared. The illustrations which appear in this revision are the work of Kathy Simpson. I feel that they add considerably to the effectiveness of the book. Her knowledge, talent, and dedication are gratefully acknowledged.

Arcata   JPS

Literature Cited:

Cronquist, A. 1968. The evolution and classification of flowering plants. Houghton-Mifflin Co. Boston. 396 pp.

Cronquist A., A. Takhtajan, & W. Zimmermann. 1966. On the higher taxa of Embryobionta. Taxon 15:129-134

Hutchinson, J. 1967. Key to the families of flowering plants of the world. Clarendon Press. Oxford. 117 pp.

Hutchinson, J. 1969. Evolution and phylogeny of flowering plants. Academic Press. New York. 717 pp.

Hutchinson J. 1973. The families of flowering plants. 3rd ed. Oxford at the Clarendon Press. 968 pp.

Lawrence, G. H. M. 1951. Taxonomy of vascular plants. Macmillan Publ. Co. New York. 823 pp.

Melchior, H. & E. Werdermann. 1954. Syllabus der Pflanzenfamilien. 12th ed. 2 vols. Bebrüder Borntraeger. Berlin.

Willis, J. C. 1973. A dictionary of the flowering plants and ferns. 8th ed. Revised by H. K. Airy Shaw. Cambridge Univ. Press. 1245 pp.

Abbreviations:

A - androecium
ach - achene
Afr - Africa
agg - aggregate
Am - America
As - Asia
AS - Anglo-Saxon
Aust - Australia
ber - berry
C - corolla
cap - capsule
cary - caryopsis
cosmo - cosmopolitan
dim - diminutive
drp - drupe
drplt - druplet
Euras - Eurasian
flw - flower
fol - follicle
Fr - French
frt - fruit
G - gynoecium
Gk - Greek

H - herb
ICBN - Int. Code Botanical Nomenclature
IE - Indo-European
Ind - India
K - calyx
L - Latin
leg - legume
ME - Middle English
Medit - Mediterranean
Mex - Mexico
mult - multiple
NC - northcentral
nutl - nutlet

NW - New World
OE - Old English
OW - Old World
pantrop - pantropical
pom - pome
p.p. - in part
pyx - pyxis
S - shrub
sam - samara
schizo - schizocarp
sil - silicle
s. l. - in the broad sense
slq - silique
Sp - Spanish
s. s. - in the strict sense
subarc - subarctic
subtrop - subtropical
syc - syconium
T - tree
temp - temperate
trop - tropical
utr - utricle
V - vine
z - zygomorphic

± - more or less
[ ] - rarely
= C - same as number of petals
=1-2K - same number or twice the number of sepals

1- The Taxonomic Hierarchy And Scientific Names

In the strictest sense, the classification of plants involves placing them in a series of categories or logical classes which have been ordered or arranged to show relationships to one another. The names and sequence of these categories are set by the "International Code of Botanical Nomenclature" (ICBN). When these categories are so arrayed, they constitute the **taxonomic hierarchy,** the list of categories into which plants are classified. Any one of these groups, at any level, may be referred to as a **taxon** (plural, **taxa**). The principal ranks of the taxonomic hierarchy and their standard endings, where applicable, are shown below.

Several other taxonomic ranks may be intercalated as needed. A more complete listing of these categories, arranged in descending order with standard endings is: division (**-phyta**), subdivision (**-phytina**), class (**-opsida**), subclass (**-idae**), order (**-ales**), suborder (**-ineae**), family (**-aceae**), subfamily (**-oideae**), tribe (**-eae**), subtribe (**-inae**), genus, subgenus, section, subsection, series, subseries, species, subspecies, variety, subvariety, forma, and subforma. The taxa at the rank of genus and below do not have standard endings. There are, however, considerations of technical word endings which must be satisfied. The elements of the scientific name must agree in gender, case, and number.

While all of the families of vascular plants have the standard -aceae ending, eight of them have equally correct alternative names.

Table 1. The categories of the taxonomic hierarchy and their standard endings.

Rank	Standard Ending	Example
Division	-phyta	Magnoliophyta
Class	-opsida	Liliopsida
Subclass	-idae	Liliidae
Order	-ales	Cyperales
Family	-aceae	Poaceae
Genus		*Triticum*
Species*		*aestivum*

*See text for a more accurate explanation of this term.

Table 2. The eight families with two correct names.

Traditional Family Name	Acceptable Alternative
Compositae	Asteraceae
Cruciferae	Brassicaceae
Gramineae	Poaceae
Guttiferae	Clusiaceae
Labiatae	Lamiaceae
Leguminosae	Fabaceae
Palmae	Arecaceae
Umbelliferae	Apiaceae

These alternatives are permitted because they predate the modern conventions of standardized endings. I prefer the traditional names for these families and they are used in this text.

THE COMPONENTS OF SCIENTIFIC NAMES

If we examine the botanical works of the 15th and 16th centuries, we see that the name of a plant was often a rather lengthy series of descriptive words, usually in Latin. These **phrase names** or **polynomials** became increasingly awkward. The discovery of a new kind of plant required that the existing polynomial be slightly modified so that the new plant could be distinguished from the old one. Consider the following example.

Convolvulus folio Altheae (Clusius, 1576)

Convolvulus argenteus Althaea folio (Bauhin, 1623)

Convolvulus argenteus foliis ovatis divisis basi truncatis: laciniis intermediis duplo longioribus (Linnaeus, 1738)

A system of nomenclature was developed which has now completely replaced these cumbersome phrase names. At the base of this new system is the principle that each plant is given a two - element name or **binomial.** At first, the use of binomials in scientific and popular writings was sporadic. The event which, perhaps more than any other, assured their permanent use in scientific writings was the adoption of the binomial system by Carolus Linnaeus (1707-forever) in his **Species Plantarum.** This work now stands as the starting point for the formal nomenclature of the higher plants. On p. 863, we find the following entry:

11. *Conyza foliis ovalibus in-   hirsuta tegerrimis scabris subtus hirsutis.*

In this example, the "11" indicates the number Linnaeus assigned to this particular kind of *Conyza; Conyza* is the **genus** or generic name; the remainder of the phrase is the diagnosis; and the *hirsuta* in the margin is the **nomen triviale** or **trivial name.**

Today we find the **Binomial System of Nomenclature** in an even more simplified form. The two elements of the binomial which make up the scientific name are derived from the taxonomic hierarchy discussed earlier. The first element of the scientific name is called the **genus** (plural, **genera**), just as in the days of Linnaeus. The second element is usually referred to as the species (singular and plural have same spelling), but this is incorrect. The second element should be called the **specific epithet.** It is the genus and the specific epithet together which form the **species** name. The binomial is the species name, not merely the second element. This more rigorous use of "species" is often ignored by student and professional botanist alike in informal conversations and writing.

The binomial, for reasons of completeness and accuracy, is followed by the abbreviation of the name of the person or persons who first applied that name to the plant. For example, the "L." in *Agrostis alba* L. stands for Linnaeus. These abbreviations constitute the **authority.** It may appear somewhat complicated by two additions, **"ex"** and **"in".** The Latin word "ex" means "from" or "according to". It is used to connect the names of two authors of which the second validly published a name proposed, but not validly published, by the first, as in *Gossypium tomentosum* Nutt. ex Seem. The word **"in"** is used to connect the names of two authors when the first supplied a name with a description in a work actually published by the second author, as in *Viburnum ternatum* Rehder in Sargent.

It is sometimes necessary to transfer a specific epithet from one genus to another. In 1803 the French botanist Andre Michaux discovered a plant which he described as *Andropogon ambiguus*. In 1888 three other botanists (Britton, Sterns, and Poggenburg), studying the same group of grasses, found that Michaux's plant was better accomodated in the genus *Gymnopogon*. The correct name for that plant would now be written *Gymnopogon ambiguus* (Michx.) B. S. P. The name in parentheses is that of the person who first applied that specific epithet to the plant; followed by the abbreviation of the person who transferred the epithet to another genus.

There are a few simple rules which must be followed in writing scientific names. The genus is always capitalized. The specific epithet should not be capitalized. It may be if it incorporates the name of a person, as in *Agropyron Smithii*, or if it was once used as a generic name, as in *Acer Negundo*. Even under these circumstances, capitalization is discouraged by the ICBN. Both elements of the scientific name should be underlined when handwritten or typed; in italics or bold-

face when in print. The authority is always capitalized, but is not underlined.

It is often useful to recognize taxonomic categories below the species level (**subspecific categories**). The three most widely used are the **subspecies, variety,** and **forma.** Each has its own authority, as in *Agrostis exarata* Trin. var. *monolepis* (Torr.) Hitchc. If the subspecific name repeats the epithet, the authority is not repeated, as in *Lomatium triternatum* (Pursh) Coult. & Rose var. *triternatum*.

It is sometimes not possible, nor perhaps necessary to include the specific epithet. In such instances the abbreviations **sp.** (plural, **spp.**) may be used, as in *Agrostis* sp. or *Quercus* spp. The abbreviations are not underlined.

Hybrids are also accorded formal designation in scientific names. Interspecific hybrids, involving the crossing of two or more species within the same genus, may be designated by a formula, as in *Salix* X *capreola* Kerner ex Andersson = *Salix aurita* L. X S. *caprea* L. The "X" between the generic name and the epithet indicates the interspecific origin of the taxon. In a more abbreviated form, the hybrid may be cited as *Salix* X *capreola* Kerner ex Andersson. In a similar fashion, intergeneric hybrids involving crosses at the genus level may be indicated by placing the "X" before the generic name, as in X *Elyhordeum* = *Elymus* X *Hordeum*, or simply X *Elyhordeum*.

THE ORIGIN OF GENERIC NAMES AND SPECIFIC EPITHETS

Most of the words which make up scientific names are derived from Latin or Greek, although there is no requirement that they must be. Modern and even nonsensical words have been used. For technical purposes, the elements of the binomial are treated as Latin, no matter what their source.

Some students have the deep-seated belief that these names are really composed by selecting letters at random from a boiling pot of vowels and consonants. The successful taxonomist is the one who can come up with a truly unpronounceable combination of letters. This reveals a lack of scholarship. A rudimentary knowledge of etymology is very helpful in understanding the composition of scientific names. In most of them are roots which often describe some feature of the plant, incorporate the name of some famous person or which perpetuate a classical name for the plant. Examples are shown in Table 3.

PRONUNCIATION OF SCIENTIFIC NAMES

The "International Code of Botanical

Table 3. The origin of generic names and specific epithets.

Name	Source or Meaning
A) Commemorative Names	
Blighia	William Bligh
Torreya	John Torrey
Lewisia	Meriwether Lewis
nuttallii	Thomas Nuttall
B) Classical or Aboriginal Names	
Agrostis	grass
Fagus	beech
mays	corn
cepa	onion
C) Geographical Names	
anglicus	of England
gallicus	of France
thapsus	of Thapsus, Italy
D) Growth Form	
arboreus	tree
repens	creeping
scandens	climbing
E) Habitat	
arenarius	growing in sand
campestris	of fields
fluviatilis	of rivers
F) Morphological feature	
Penstemon	5 stamens
Sanguinaria	red latex
amabilis	lovely
campanulatus	bell-shaped
foetidus	foul-smelling

Nomenclature" states that scientific names are treated as Latin words, regardless of their origin. A few more scholastically inclined individuals feel, therefore, that we must follow strict Latin rules of vowel and consonant sounds and accenting. It is particularly among such persons that we find the notion that there is a single correct way to pronounce a scientific name. One difficulty with this view is that there are traditional English, reformed academic, and Church Latin versions of pronunciation. At the other extreme is the view that attempting most of the syllables and perhaps adding a few that are not in the word to demonstrate creativity, is close enough. To do more is thought to be a display of excessive pedantry.

Most American botanists generally pronounce scientific names as if they were English words. Some follow Latin rules for placing the accent; many of us do not. What "sounds right" may well be determined more by the manner in which a former botany instructor pronounced these words than in following a particular set of phonetic rules. I have attempted to present below a basic guide to Americanized vowel, consonant, and diphthong sounds, along with some more formal rules for accenting. The student interested in a rigorous explanation of Latin grammar should consult Stearn (1966).

1) The letters of the Latin alphabet are much like ours, except that the letters J, U and W did not occur in the classical alphabet of the time of Cicero. Some scholars insist, for instance, that *Castilleja* should be spelled *Castilleia*.

2) Pronunciation is based upon the sounds of the letters, the nature of the vowels or their equivalents, and the place of the stress or accent.

3) Separate the scientific name into its component syllables. Each syllable will contain a vowel or diphthong, the double vowel combinations ae, au, ei, eu, oe, and ui.

4) Pronounce all of the syllables. *Ribes* is pronounced "rī-bēs", not "rībs".

5. Final vowels, with the exception of "a", are long. Two vowels together that do not form a diphthong, as the -ia in *quinquefolia*, are sounded separately, "quin-que-fo-li-ah".

6) The diphthongs ae and oe have the sound "ē"; au has the sound of " aw ", as in "awful"; ei usually has the sound "ī"; eu has the sound of "u" in "neuter"; and ui has the ui-sound, as in "ruin".

7) The oi in the Latinized Greek ending oides is pronounced as a diphthong by most American botanists. As such it would have the sound of the oi in "oil." Technically, however, the double vowels do not constitute a diphthong so that -oides should be pronounced "-o-e-des."

8) A single consonant is placed with the following vowel, as in "pa-ter". Double consonants are separated, as in "am-mi". If there are two or more different consonants, the first one is usually put with the preceding vowel, as in "an-gli-cus".

9) The consonants b, d, f, h, l, m, n, p, qu, and z are pronounced the same in Latin and English.

10) The consonants c and g are soft (have the sounds "s" and "j"), if they are followed by ae, e, i, oe, or y. Otherwise, c is pronounced as a "k" and the g is also hard, as in "go". The s is always as in "so", not a "z". An initial x is pronounced as a "z", not "ek-z". *Xanthium* is "zan-thi-um", not "ek-zan-thi-um".

11) The first letter is silent in words beginning with cn, ct, gn, mn, pn, ps, pt, and tm.

12) Accenting the proper syllable is not always easily done. Several standard references have the acute (´) and grave (`) accent marks given. Please note that these accent marks are given for the convenience of the reader and are not actually parts of the scientific name. If you must attempt to determine the proper syllable to accent, it is helpful to remember that words of two syllables are always accented on the first syllable. In words of three or more syllables, the last syllable is never accented. The stress must fall on the next to the last syllable (penult or penulti-

mate), as in "ar-vén-sis", or the third from the last syllable (antepenult or antepenultimate), as in "án-gli-cus". No matter how long the word, the accent cannot be to the left of the antepenult, as in "no-ve-bo-ra-cén-sis" or "phi-la-dél-phi-cus". Accent the penult if it ends in a consonant or a diphthong. The penult is also accented if it ends in a long vowel. It may be necessary to consult a Latin dictionary at this point.

13) Commemorative names present a special problem because proper accenting can render the base name unrecognizable. The epithet *jamesii* should be pronounced "jă-mē-sē-ī", rather than "jāmes-ē-ī". Many botanists ignore the rules in such cases. Stearn (1966) recommends pronouncing foreign commemorative names as close to the original name as possible, but with Latin endings. Thus the unpronounceable orchid genus *Warszewiczella* is rendered "văr-shě-vĭ-chĕl-lă".

THE ADVANTAGES AND DISADVANTAGES OF SCIENTIFIC AND COMMON NAMES

Although scientific names may cause the beginning student some discomfort, they are well worth the effort. Their advantages to the biological sciences are rather obvious. There is a single valid, universally recognized scientific name for each plant. Because they are used by botanists all over the world, scientific names facilitate the free transfer of ideas and information.

The same name may not be used for more than one kind of plant. Once a binomial has been given to a particular species, that combination may not be used for any other plant, even if that combination is later rejected.

Scientific names are given to plants according to a very rigid set of regulations, the international code discussed earlier. This insures stability and accuracy in the application of names. These rules are reviewed regularly at the International Botanical Congresses.

Unlike many common names, scientific names have "information content". Inherent in the name of a plant is the idea of relationship. According to one very prominent school of thought, the binomial corresponds to a population or series of populations which occur naturally and which are genetically discrete. In other words, we are applying names to units which are genetically defined and which have biological reality. The various species which constitute the genus are related. This means that if you know the characters of one species that belongs to a genus, then you may predict with a high degree of confidence the characteristics of other species of that same genus, even though you may have never seen them. These ideas of information content and predictability are very important features.

There are, unfortunately, some defects in scientific names. They are not widely used by the general public. They can be difficult to pronounce. While I can accept these criticisms, it is worth noting that such widely used and easily pronounced common names as aster, rhododendron, magnolia, fuchsia, chrysanthemum, petunia, and begonia are really generic names. Other scientific names may be more obscure, but they are often no more difficult to pronounce. They are simply unfamiliar. Even though scientific names have elegant Latin or Greek roots, they should be pronounced as though they were English words.

Although all taxonomists accept the principle that a plant must have only one scientific name, we do sometimes haggle endlessly over what that name should be. This is a serious problem for the uninitiated student who wanders from reference to reference, only to find conflicting treatments of the same taxa. In that taxonomy is self-correcting and is hopefully evolving, these criticisms will not be easily satisfied.

While I am rather committed to the efficacy of scientific names, it is foolish to maintain that common names for plants have no value. They are the only names known to most people. The names are often simple and easy to remember. They may be descriptive, colorful, pleasing to the ear, and easy to pronounce.

The disadvantages of common names far outweigh these advantages. The same plant may have more than one common name. A common lawn weed, usually called "broad-leaved plantain", has almost fifty other common names in English. We may compound this problem by noting that each language has its own set of common names. The same common name may be used for a wide variety of unrelated plants. "Laurel" is applied to the plants of at least five different families. To learn the common name of a plant is no mean task!

Many common names are confusing. The pineapple is not a pine, nor is it related to the apple. Kentucky bluegrass if not a blue grass, nor is it from Kentucky. Such names as "welcome home husband, no matter how drunk ye be", "kiss me over the garden gate", "Jack, go to bed at noon", "ramping fumitory", "spotted arsemart", and "stinking orach" make it difficult to maintain that common names have clarity and brevity.

Common names are not applied according to any logical or consistent system of rules. Although a common mame may often be associated with a particular species, it is difficult in many cases to argue that it is **the** correct common name.

One final defect in common names is particularly serious. Man has used only a small portion of the half million or so kinds of plants. Most plants simply do not have common names. This seems to be most distressing to the authors of popular guides to plants and to the staffs of some government agencies. They have attempted to compensate for this lack of common names by conjuring them up. This is usually done by translating the scientific name into English. The advantage of Milo Baker's cryptantha over *Cryptantha milobakeri* is not immediately apparent.

THE INTERNATIONAL CODE OF BOTANICAL NOMENCLATURE

The reality of a universally accepted code which would govern the application of scientific names to plants is quite recent. For centuries each botanist worked in his own country, creating his own names and rules, quite unaware of similar activity by scientists in other countries. The beginnings of a modern code occurred at the First International Botanical Congress held in Paris in 1867. Serious problems developed immediately. There followed a long period of internecine warfare, pursued with almost religious fervor. Competing sets of rules were created. But, out of all this came the elements of the present International Code of Botanical Nomenclature. Today the ICBN is in an era of stability. The last two or three editions have been essentially the same, differing primarily in the correction of errors and in the refining of the language of certain rules.

The present ICBN rests on a series of principles. The first of these is that botanical nomenclature is independent of zoological nomenclature. One practical result of this is that the same names may be used for both a plant and an animal. *Sida* is the genus of a cotton relative and a crustacean; *Daubentonia* is a type of bean and a primate.

Another very important principle is that the name of a plant is permanently associated with a particular reference specimen, the **nomenclatural type.** This is the specimen which the author of the name declares in his original publication. It is a purely nomenclatural device, not a biological model. Much confusion has been generated by those who look upon the nomenclatural type as the archetype of Aristotelian philosophy. It is not the standard against which unknowns are measured. The type need not even be typical. There are several different kinds of types. The most important of these is the **holotype,** the single specimen designated by the author from the original collection of his new entity.

Duplicates of the holotype, if they exist, are called **isotypes.**

Of prime importance in determining the correct scientific name is the principle of priority of publication. The first published name of a plant is accepted as its correct name. There is an escape clause, however. There are instances when well known and long accepted names would have to be rejected in favor of an earlier more obscure name, if the rules of priority were strictly applied. There is a mechanism for conserving such widely used family and generic names over those which have technical priority. Specific epithets cannot be conserved, however.

Each taxon can have only one correct name. The only exceptions to this are the eight family names already mentioned.

The bulk of the ICBN is made up of specific rules and recommendations. Some of the more important of these are:

1) Valid publication of the names of ferns, gymnosperms, and flowering plants begins 1 May 1753, with the publication of the **Species Plantarum** by Linnaeus.
2) Family names are formed by adding aceae to the stem of an included genus, as in *Geranium* and Geraniaceae.
3) The names of genera may be taken from any source, and may even be completely arbitrary, as in *Muilla,* which is *Allium* spelled backwards.
4) The name of a species is a binary combination consisting of the genus followed by a specific epithet.
5) **Effective publication** of a new name requires that printed matter be distributed to the public or, at least, to botanical institutions with libraries. Oral presentations, microfilm, mimeographed handouts and distribution of annotated specimens are not sufficient.
6) For **valid publication** of a new name, the author must either prepare a Latin description or provide a reference to a previously published description.

7) A name cannot be rejected because someone feels that it is inappropriate, because another might be better, or because the name has lost its original meaning. For instance, the name *Scilla peruviana* cannot be rejected simply because the plants do not grow in Peru.
8) A name must be rejected if it is spelled exactly like the name of a previously and validly published name of a taxonomic group of the same rank, based upon a different nomenclatural type. A *stragalus rhizanthus* Boiss, published in 1843 is rejected in favor of the validly published name *Astralagus rhizanthus* Royle which appeared in 1835.
9) The original spelling of a name should be retained, except for the correction of typographic or **orthographic errors.** The Linnaean genera *Amaranthus* and *Gleditsia*, for instance, cannot be altered to *Amarantus* and *Gleditschia*, although these spellings are philologically preferable.

USEFUL REFERENCES:

Gray Herbarium Card Index. 1894-. Harvard University, Cambridge, Mass. An index of all new names and combinations of western hemisphere vascular plants. Present size about 285,000 cards.

Index Kewensis Plantarum Phanerogamarum. 1893-. Claredon Press, Oxford. 2 vols. + 14 supplements comprising an index of all new names and combinations of seed plants published from the time of Linnaeus to the present.

McVaugh, R., R. Ross, & F. A. Stafleu. 1968. An Annotated Glossary of Botanical Nomenclature. Regnum Vegetabile Vol. 56. International Bureau for Plant Taxonomy and Nomenclature of the International Association for Plant Taxonomy. Utrecht, Netherlands. 31 pp.

Stafleu, F. A. (Editor). 1972. International Code of Botanical Nomenclature. Adopted by the 11th International Botanical Congress, Seattle, Washington, August 1969. A. Oosthoek's Uitgeversmaatschappij N.V., Utrecht.

Stern, W. T. 1966. Botanical Latin. Nelson & Son. London. 566 pp.

2- The Ferns and Their Allies

The ferns and their allies comprise the Psilophyta, Rhyniophyta, Lycopodiophyta, Equisetophyta, and Polypodiophyta. The plants of these divisions, along with those of the Pinophyta and Magnoliophyta, are referred to as the vascular plants. The name is based upon the presence of specialized conducting tissues, the xylem and phloem, in the sporophyte, the commonly observed generation in the life cycle. The most primitive vascular plants, those of the Psilophyta and Rhyniophyta, are not clearly differentiated into root, stem, and leaf. This distinction is rather easily seen in the remaining divisions, except in certain specialized groups such as the duckweeds and water ferns. The rhyniophytes, known only from fossil remains, are not treated here.

DIVISION PSILOPHYTA

This small division, as it is traditionally treated, consists of two extant genera. Fossil representatives, belonging to the Psilophytales, are usually considered the most primitive vascular plants. Recent work has suggested that such an interpretation may not be correct.

Psilotaceae
Whisk Fern Family

Epiphytic perennial herbs. Rootless, but the rhizome with rhizoids. Aerial stem erect, green, much-branched. Leaves or leaf-like emergences minute. Sporangia comparatively large, 3-chambered, homosporous.

2/10; tropical and subtropical. *Psilotum* is found in the southern U.S. None is of economic importance.

Genera:

Psilotum (3)-whisk fern; N. America, Japan, Australia, and New Zealand

Tmesipteris (7)-Australia, New Zealand, and Polynesia

DIVISION LYCOPODIOPHYTA

All of the plants of the Lycopodiophyta have true roots, stems, and leaves. Many have rhizomes. The aerial stem is either erect or prostrate. The leaves are often small, 1-veined, awl-shaped, and spirally arranged. All produce sporangia on specialized fertile leaves, the **sporophylls.** Some lycopods are homosporous; others are heterosporous.

Lycopodiaceae
Club Moss Family

Terrestrial or epiphytic moss-like herbs. Stems erect or prostrate, freely-branched. Leaves small, numerous, 1-veined, in spirals, whorls, or pairs. Sporangia on the upper side of the sporophylls, these usually aggregated into terminal strobili; homosporous.

2/450; tropical and temperate. Several species of *Lycopodium* are native to the U.S. None is of any real economic importance. The spores were once used as flash powder.

Selected Genus:

Lycopodium (450)-club moss, ground pine

Selaginellaceae
Spike Moss Family

Low erect or spreading perennial herbs. Stems freely dichotomously branched. Leaves small, 1-veined, of two kinds, spirally inserted or 4-ranked, a minute flap **(ligule)** at the base of each leaf. Heterosporous, the sporophylls either aggregated into a definite strobilus or not grouped and resembling the sterile leaves.

1/700; principally tropical, some temperate. Many species are native to the U.S. Several are cultivated as ornamentals.

Genus:

Selaginella (700)-spike moss

Isoetaceae
Quill Wort Family

Small perennial aquatic or emergent herbs of grass-like habit. Stems short, erect. Leaves quill-like, arising in clumps from the corm-like stem; blades 1-veined with ligule. Outer leaves with megasporangia sunken in the leaf bases; inner leaves with microsporangia.

2/77; cosmopolitan. Several species of *Isoetes* occur in the U.S. None is of economic importance.

Genera:

Isoetes (75)-quill wort

Stylites (2)-restricted to the Peruvian Andes

DIVISION EQUISETOPHYTA

This division is represented today by a single genus. Fossil evidence suggests that the group was once much more extensive and diverse. The group probably reached its height during the Upper Carboniferous. One prominent extinct member is *Calamites*, which closely resembles the living horsetails, but it grew to the size of small trees of 10-15 m. Today's horsetails are much smaller, although *Equisetum* X *schafneri* of Central America may be over 6 m tall.

Equisetaceae
Horsetail Family

Annual or perennial herbs with erect, distinctly jointed, fluted, silicaceous stems which arise from an underground rhizome. The stems are unbranched or with whorls of branches which are easily mistaken for leaves. Leaves reduced to a whorl of fused scales. Strobili terminal, borne either on ordinary green stems or on specialized fertile stems which lack chlorophyll. Strobili consisting of a series of peltate sporangia scales attached to a central axis.

1/23; cosmopolitan, except Australasia; primarily of wet sites. Several species are native to the U.S. None is of any current economic importance. The silica-impregnated stems were used by early settlers as scouring devices, hence one of the

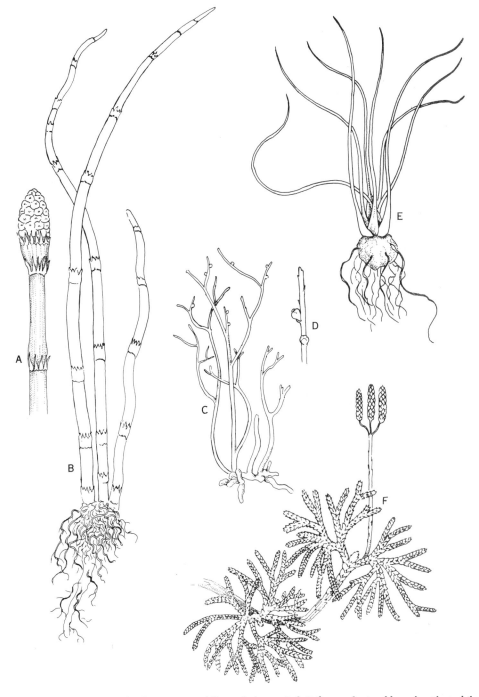

Fig. 1. **Lower vascular plants.** a & b, *Equisetum* strobilus and plant; c & d, *Psilotum* plant and branch with strobilus; e, *Isoetes;* f, *Lycopodium.*

common names.

Genus:

Equisetum (23)-horsetail, scouring rush, snake grass

DIVISION POLYPODIOPHYTA

This division contains the true ferns and their extinct allies. Almost all of the living ferns have true roots, stems, and leaves. In most temperate zone species, the stem is reduced to an underground creeping rhizome which bears adventitious roots. In certain tropical species, the "tree ferns", the stems are well-developed and the plants tree-like. The leaves of most ferns are usually quite distinctive, often variously dissected, and show a characteristic unrolling of the young leaves, **circinate vernation.** The fern leaf is called a **frond;** the leaf stalk a **stipe.** The blade may be simple or more commonly compound. The primary divisions of the blade are referred to as **pinnae** (singular, **pinna**). Each pinna may be further subdivided into still smaller segments, the **pinnules.** The leaves are **macrophyllous,** with a complex vascular system, **leaf gaps,** and **leaf traces.** Sporangia are borne on the undersurface of ordinary fronds or on very specialized fertile fronds or segments thereof. The sporangia themselves may occur singly or in clusters called **sori** (singular, **sorus**) or may be in **synangia,** fused sori. The sorus may be covered by a flap of leaf tissue, the **indusium** or it may be without this protection **(naked** or **exindusiate).** The majority of ferns are homosporous, although heterospory is known in some of the more specialized aquatic families.

The classification of the Polypodiophyta remains controversial. Much of the work that has been done in the past few decades consists of attempts to redefine the limits of families and genera. Noted authorities are still in major disagreement with one another. The system used here is that of Crabbe, Jermy, and Mickel (Fern Gazette 11(2 & 3):141-162).

Ophioglossaceae
Grape Fern Family

Perennial terrestrial herbs. Leaves 1 to a few, differentiated into a sterile blade and a distinctly fertile segment which is spike-like or branched. Plants homosporous.

4/70; widely distributed in temperate and tropical areas. *Ophioglossum* and *Botrychium* are represented in the U.S. by several species. None is of economic importance.

Selected Genera:

Ophioglossum (30-50)-adder's tongue fern

Botrychium (40)-grape fern

Osmundaceae
Osmunda Family

Terrestrial herbs. Leaves (in ours) pinnately compound with a clear distinction between the vegetative and fertile pinnae. Sporangia on the upper and lower sides of the modified segments. Sori naked.

3/19; tropical and temperate. A few species of *Osmunda* occur in the U.S. Some are ornamentals.

Selected Genus:

Osmunda (10)-royal fern, cinnamon fern, interrupted fern

Schizaeaceae
Curly Grass Family

Terrestrial herbs. Sporangia comparatively large, sessile, and borne on specialized pinnae or parts thereof. Mostly tropical and subtropical.

Selected Genera:

Schizaea (30)-curly grass; Alaska

Lygodium (40)-a climbing fern of the southern states

Adiantaceae

Terrestrial herbs, some of xerophytic sites. Sori often covered by a reflexed leaflet margin, indusia absent.

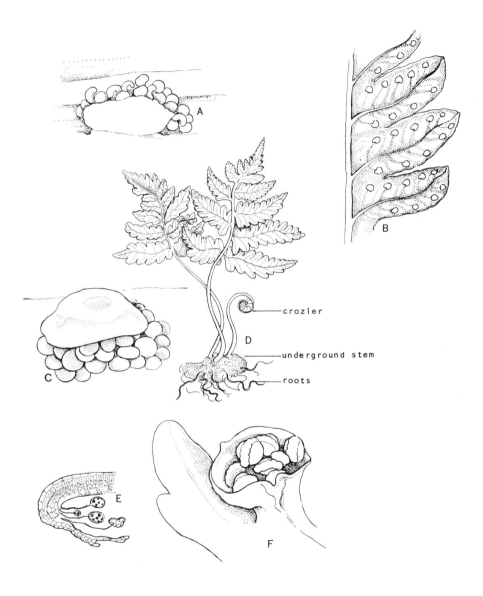

Fig. 2. **Fern morphology.** a, sorus showing side attachment; b, segment of frond showing sori; c, sorus with peltate attachment; d, fern plant (arrow indicates crozier); e, pinnule showing folded margin forming false indusium; f, sorus showing cup-shaped indusium.

Selected Genera:

Onychium (6)-cliff brake

Cryptogramma (4)-parsley fern, rock-brake

Cheilanthes (180)-lace fern, lip fern, cloak fern

Aspidotis (2)-cliff brake

Pellaea (80)-cliff brake

Pityrogramma (40)-gold fern, goldenback

Adiantum (200)-maidenhair fern

Pteris (250)-brake fern

Hymenophyllaceae
Filmy Fern Family

Delicate herbs, the fronds only one cell thick. Sporangia at the ends of veins, indusiate. Primarily tropical and temperate, with very limited distribution in the U.S.

Selected Genera:

Trichomanes (25)-filmy fern; southern states

Hymenophyllum (25)-filmy fern; South Carolina only

Polypodiaceae
Polypody Family

Typically epiphytes. Sori without indusia. Cosmopolitan. The family, as it was traditionally delimited, was a huge heterogeneous assemblage of ferns.

Selected Genera:

Polypodium (75)-polypody

Platycerium (17)-staghorn fern; widely cultivated, but none is native to the U.S.

Cyatheaceae
Tree Fern Family

Tree-like plants to 6 m tall. Leaves to 6 m long. Sori on veins or in the forks of veins. Indusiate or exindusiate. Tropical and subtropical; none is native to the U.S., although they are widely cultivated.

Selected Genus:

Cyathea (including *Hemitelia* and *Alsophila*) (600)-tree fern

Dennstaedtiaceae

Large terrestrial ferns with hairy rhizomes. Sori marginal or submarginal; indusiate. Cosmopolitan.

Selected Genus:

Pteridium (1)-*P. aquilinum* is the bracken fern, a widespread toxic plant

Aspleniaceae

Epiphytic or terrestrial, many of rocky sites. Sori on the veins or at the ends of veins. Indusia usually present, rounded to reniform, peltate.

Selected Genera:

Asplenium (650)-spleenwort

Camptosorus (2)-*C. rhizophyllus* is the walking fern

Matteuccia (3)-*M. struthiopteris* is the ostrich fern

Onochlea (1)-*O. sensibilis* is the senstivie fern

Athyrium (180)-*A. filix-femina* is the lady fern

Cystopteris (18)-bladder fern

Woodsia (40)-woodsia

Polystichum (135)-holly ferns

Dryopteris (150)-oak fern, marsh fern, male fern

Thelypteris (4)-often placed in its own family

Davalliaceae

Mostly epiphytes. Sori terminal or on the veins.

Selected Genus:

Nephrolepis (30)-sword fern; *N. exaltata* is the widely cultivated Boston fern

Fig. 3. **Terrestrial and epiphytic ferns.** a, *Polystichum* frond; b & c, *Adiantum* pinnule and frond segment; d, *Ophioglossum* showing fertile and sterile segments; e, *Polypodium* frond; f, *Osmunda* showing fertile and sterile pinnae.

Blechnaceae
Deer Fern Family

Terrestrial herbs. Sori on veins parallel to the midrib of the leaflet. Indusia present. Cosmopolitan.

Selected Genera:

Blechnum (220)-deer fern

Woodwardia (12)-chain fern; plants often quite large

Marsileaceae
Water Clover Family

Small aquatic or semiaquatic rhizomatous herbs of muddy lake and pond margins. Leaves either bladeless or with four leaflets in our species. Sporangia borne in small bean-like structures, the **sporocarps,** at the base of the leaf stalks. Plants heterosporous.

3/70; temperate and tropical. *Marsilea* and *Pilularia* are native to the U.S. Of limited importance as ornamentals.

Genera:

Marsilea (60)-water clover; 4 leaflets

Regnellidium (1)-2 leaflets; restricted to Brazil

Pilularia (6)-pillwort; no leaflets, plants grass-like

Salviniaceae
Salvinia Family

Free-floating aquatics. Stems delicate. Leaves 3 per node, two floating on the surface and one finely divided and root-like leaf submerged. Roots absent. Heterosporous, sporangia enclosed in sporocarps from the base of submerged leaves.

1/10; tropical and temperate. Escaped in the eastern U.S.

Genus:

Salvinia (10)-salvinia

Azollaceae
Mosquito Fern Family

Free-floating aquatics. Stems delicate. Leaves paired, bilobed. Roots adventitious, hanging into the water. Heterosporous, sporangia enclosed in sporocarps on the ventral lobes of the first leaves of the branch.

1/6; tropical and subtropical. The family is traditionally considered part of the Salviniaceae.

Genus:

Azolla (6)-mosquito fern; native to the U.S.

Fig. 4. **Aquatic ferns.** a, *Salvinia;* b, *Azolla;* c, *Marsilea;* d, *Marsilea* sporocarps; e, *Regnellidium;* f, *Pilularia.*

3- The Pinophyta (The Gymnosperms)

This division is a diverse assemblage of plants which represent different evolutionary lines. Recent taxonomic treatments of the gymnosperms have tended to emphasize these differences more than did the classical works. Other modern systems of classification divide them among three or four divisions. Relationships among the groups, at whatever level they are recognized, remain most controversial.

Three groups, the seed ferns, fossil cycads, and the Cordaitales, are known only from fossil material. They are excluded from this treatment. The living representatives of the Pinophyta are the conifers, the cycads, maidenhair tree *(Ginkgo),* and the anomalous genera *Gnetum, Ephedra,* and *Welwitschia.* The last three genera are of particularly uncertain affinities. Workers of the 19th century placed them in the order Gnetales, more out of a sense of morphological and taxonomic desperation than anything else.

The only common group of gymnosperms in North America is the conifers. These are the familiar cone-bearing trees of the forested areas of the continent. Most are well-developed trees, although a few are shrubby. None is herbaceous. Many of the conifers show two different kinds of branches. The "long shoots" are capable of extended linear growth and may achieve great size. The "short shoots" or "spurs" are stunted branches which may bear a few to many leaves. In those plants with both kinds of shoots, the leaves are usually restricted to the spur shoots, as in *Pinus, Cedrus,* and *Larix.* The reproductive structures of the conifers are usually borne in strobili of varying size and complexity. The staminate strobili are often catkin-like and somewhat inconspicuous. The ovulate strobilus is typically woody and is the familiar "cone" that most of us associate with the evergreen trees. In some instances the cone is not woody, but berry-like. In still other species, the ovules may not be borne in fully developed strobili, but may occur singly or in pairs. In such cases they resemble fleshy fruits.

The Pinophyta is composed of about 64 genera and 570 species, distributed through the temperate, subtropical, and tropical regions of both the New World and the Old World. Many of them are exceedingly important as timber trees and as ornamentals.

SUBDIVISION CYCADICAE

Cycadaceae
Cycad Family

Woody plants, stems usually unbranched, often resembling palms or tree ferns in their appearance. Leaves spirally inserted, pinnately compound, with a woody sheathing base. The species dioecious. Heterosporous, with both kinds of sporangia in separate well-developed cones (except in *Cycas*), which vary considerably in size. Scales of the ovulate strobilus typically with two marginal ovules. Cotyledons 2.

10/100; tropics and subtropics of the Old World and New World. *Zamia* is native to Florida. None is of real economic importance; several are ornamentals. One recent treatment divides the family into three on the basis of certain leaf characters.

Selected Genera:

Cycas (20)-sago palm

Zamia (30-40)-coontie of Florida and the West Indies

Microcycas (1)-a Cuban species

Dioon (3-5)-Mexico and Central America

Ceratozamia (4)-Mexico

SUBDIVISION PINICAE

Ginkgoaceae
Maidenhair Tree Family

Trees to 30 m high. Leaves deciduous, fan-shaped, notched, and dichotomously veined. The species dioecious. Ovules terminal, paired, resembling a fleshy fruit, and decidely malodorous at maturity. Cotyledons 2.

1/1; eastern China, but widely cultivated as an ornamental. Also valued for its oil, timber, and as a food source.

Genus:

Ginkgo (1)-*G. biloba* is the maidenhair tree

Pinaceae
Pine Family

Trees, rarely shurbs. Leaves persistent in ours (except in *Larix*), spirally arranged, acicular to linear, separate or fascicled, on long shoots or spurs. The species monoecious. Ovulate cone woody with spirally arranged imbricate scales. Bract and ovuliferous scale distinct. Ovules 2 per scale, inverted. Polycotyledonous.

10/250; northern and southern hemispheres of the Old World and New World. Six genera are native to the U.S. Many of great economic importance as a source of timber and naval stores.

Selected Genera:

Pinus (70-100)-pines

Larix (10-12)-larch, tamarack

Cedrus (4)-Atlas cedar, deodar cedar, cedar of Lebanon; widely cultivated although none is native to the U.S.

Abies (50)-fir

Picea (50)-spruce

Tsuga (15)-hemlock

Pseudotsuga (7)-*P. menziesii* is the Douglas-fir

Cupressaceae
Cedar or *Cypress Family*

Trees or shrubs, erect or prostrate. Leaves persistent, scale-like or awl-shaped, or both forms present on the same plant, often closely appressed to the branches. The species monoecious or dioecious. Ovulate cones usually woody, but with fleshy scales in *Juniperus*. Bract and ovuliferous scale fused;

ovules 2-many per scale, erect. Cotyledons 2, rarely 5 or 6.

19/130; cosmopolitan. Five genera are native to the U.S. Economically important as timber trees and ornamentals. Juniper "berries" are used to flavor gin.

Selected Genera:

Cupressus (15-20)-cypress

Calocedrus (= *Libocedrus* p.p.) (3)-*C. decurrens* is the incense cedar

Thuja (5)-arbor-vitae

Chamaecyparis (7)-Port Orford cedar, white cedar

Juniperus (60)-juniper, red cedar

Taxodiaceae
Bald Cypress Family

Medium to giant trees. Leaves persistent or deciduous, alternate (except in *Metasequoia*), linear to ovate or awl-shaped. The species monoecious. Ovulate cones woody, globose, with peltate scales in our species. Bract and ovuliferous scale completely or partially fused. Ovules 2-9 per scale, usually erect. Cotyledons 2-9.

10/16; North America, Tasmania, and eastern Asia. Three genera are native to the U.S. Economically important as timber trees and as ornamentals.

Selected Genera:

Taxodium (3)-*T. distichum* is the bald cypress of the South

Sequoia (1)-*S. sempervirens* is the redwood of Calif. and Oregon

Sequoiadendron (1)-*S. giganteum* is the bigtree, a California endemic and the largest living thing on Earth.

Metasequoia (1)-*M. glyptostroboides* is the dawn redwood

Cryptomeria (1)-Japanese cedar, an ornamental

Cunninghamia (3)-cunninghamia, widely cultivated

Sciadopitys (1)-*S. verticillata* is the umbrella pine

Taxaceae
Yew Family

Much branched trees or shrubs. Leaves persistent, spirally inserted with acicular, linear to lanceolate blades. The species dioecious, rarely monoecious. Ovules solitary, terminal, erect. Seeds more or less covered by a fleshy aril. Cotyledons 2.

5/20; northern hemisphere, Celebes, and New Caledonia. Two genera are native to the U.S. Economically important as a source of timber and ornamentals.

Selected Genera:

Taxus (10)-yew; both native and introduced species toxic

Torreya (6)-California nutmeg; another species in Florida

Podocarpaceae
Podocarpus Family

Trees and shrubs. Leaves scale-like, acicular, linear, lanceolate or ovate. The species monoecious or dioecious. Ovule terminal, solitary, often in an enveloping aril; at maturity forming a fleshy "fruit". Cotyledons 2.

6/125; mostly in the southern hemisphere. None is native to the U.S., although *Podocarpus* is widely cultivated. The timber is valuable; many are ornamentals.

Selected Genus:

Podocarpus (100)-podocarpus; widely planted ornamental trees

Araucariaceae
Araucaria Family

Trees, often quite large. Leaves persistent, alternate, broad or acicular. The species monoecious or dioecious. Ovulate cones globose. Bract and ovuliferous scale fused; ovule solitary. Cotyledons 2, rarely 4.

2/38; southern hemisphere, except Africa. None is native to the U.S., although *Araucaria* is a widely cultivated ornamental.

Genera:

Araucaria (18)-monkey puzzle, Norfolk Island pine

Agathis (20)-dammar pine, kauri

SUBDIVISION GNETICAE

Ephedraceae
Mormon Tea Family

Much branched shrubs or small trees. Stems grooved and green. Leaves opposite or whorled, fused, reduced to scales. The species usually dioecious. Staminate cones with 2-8 anthers. Ovulate cones complex with the ovules covered by bracts. Cotyledons 2.

1/40; warm and arid regions of the Old World and New World. Several species are native to the U.S. An Asiatic species is the source of ephedrine. The common name "Mormon tea" comes from its use as a beverage plant by the Latter Day Saints pioneers in the American West.

Genus:

Ephedra (40)-Mormon tea, ma huang (the source of ephedrine)

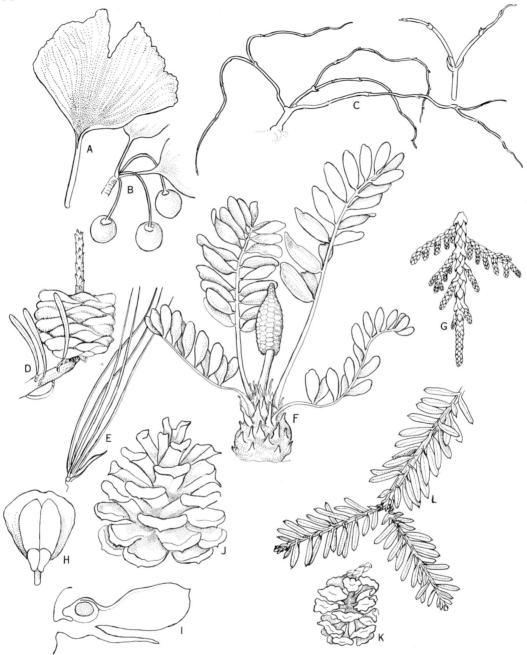

Fig. 5. **Pinophyta.** a & b, *Ginkgo* leaf and ovulate cones; c, *Ephedra;* d, *Abies* cone with disarticulated cone scales revealing central axis; e, 5-needle fascicle of *Pinus;* f, *Zamia;* g, *Thuja;* h & i, ovuliferous scale as viewed from above and from the side; j, ovulate cone of *Pinus;* k, branchlets, leaves, and cone (note pelate scales) of *Sequoia.*

4- Vegetative Morphology of the Flowering Plants

Although this chapter is designed primarily to survey the terminology describing vegetative structure of flowering plants, several of the terms apply equally well to nonflowering vascular plants.

DURATION

annual-living for one year or less

biennial-living for two years, typically flowering and fruiting the second year

caducous-applied to plant parts which fall off early or prematurely

deciduous-plants which shed all their leaves at the end of each growing season; also applied to plant parts which fall off

evergreen-remaining green during the dormant season, the plants never without some leaves

fugaceous-withering away or falling off very early

marescent-withering or fading, but remaining attached

perennial-living for three years or more

persistent-remaining attached and unwithered

HABIT

The following terms describe the general appearance or habit of plants. These growth forms should be looked upon as tendencies rather than rigid categories.

acaulescent-more or less stemless, the stem often subterranean

arborescent-tree-like, the main trunk relatively short

arboreous-trees, the trunk well-developed

caespitose (=**cespitose**)-growing in tufts, mats or clumps

caulescent-aerial stem or stems evident

clambering- spreading over undergrowth or objects, usually without the aid of twining stems or tendrils

climbing-ascending upon other plants or objects by means of special structures, such as disc-like stem tips

decumbent-stems lying upon the ground, but with their ends turned up

divaricate-extremely divergent, more or less at a right angle

divergent-broadly spreading

fruticose-shrubby, with more than one major stem

geniculate-bent sharply, as at the knee

herbs-plants with non-woody aerial stems which die back to the ground each year

lianas-woody plants with elongate, flexible, non-self-supporting stems

prostrate (=**procumbent**)-lying flat upon the ground; typically without adventitious roots

repent (=**trailing**)-stems prostrate, creeping or sprawling and often rooting at the node

resupinate-inverted because of a 180⁰ twist in a petiole or pedicel

scapose-bearing a flower or inflorescence on a leafless flowering stem

shrubs-woody perennials with more than one principal stem arising from the ground

spreading-oriented outward and more or less diverging from the point of origin

suffruticose (=**suffrutescent**)-plants woody at the base, but herbaceous above

trees-woody perennials with a single main stem or trunk

twining-coiling around plants or objects as a means of support

vine-herbaceous plants with elongate, flexible, non-self-supporting stems

TERMINOLOGY APPLIED TO THE ROOT SYSTEM

The root system of the flowering plants provides comparatively few taxonomic characters. The terminology is rather meager.

adventitious-those roots which arise from any point other than as a portion of the primary root system

aerial-those roots occurring above ground

fibrous-a root system in which all of the roots are of about the same size so that none is clearly dominant, as in many monocots

subterranean-roots occurring below the ground

tap-a root system in which one root is clearly larger than the others, as in many dicots

tuberous-a tap root which is particularly large and fleshy; not clearly delimited from the tap root

TERMINOLOGY APPLIED TO THE STEM SYSTEM

The stem system provides many useful features. Its general appearance does much to determine the habit of the plant, as discussed earlier in this chapter. Stems may be either **aerial** or **subterranean.** Aerial stems are **herbaceous** if they remain soft and produce little secondary growth or **woody** if they produce secondary growth and are covered with bark. Most woody stems also exhibit such surface features as leaf scars, stipule scars, and lenticels. Aerial stems are subject to several modifications, some of them pronounced.

cladophyll (=**phylloclad**)-a flattened leaf-like stem

stolon (=**runner**)-a horizontal stem, often rooting at the nodes, which bears ordinary foliage leaves

tendril-a twining stem, either terminal or arising from the axil of a leaf; tendrils may also be of leaf origin

thorn-a sharp-pointed stem, either simple or branched; see also spine and prickle

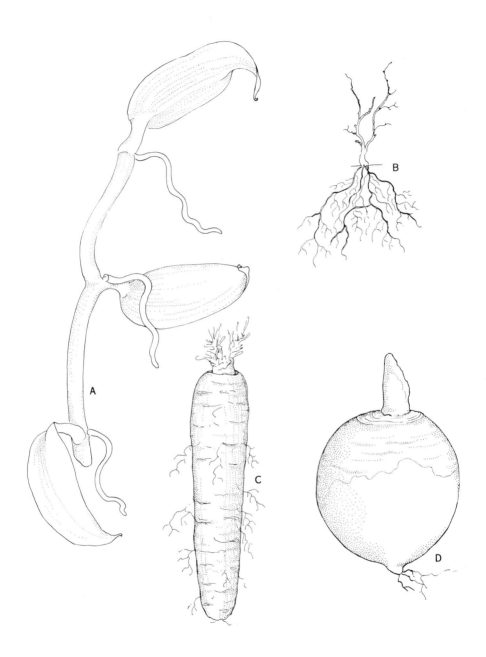

Fig. 6. **Root morphology.** a, aerial roots of an orchid; b, fibrous root system; c, tap root; d, tuberous root.

Subterranean stems because of their location are often confused with roots. The presence of nodes, internodes, leaf scars, and buds on these underground structures should assist in distinguishing them from roots. Underground stems may also be heavily modified.

bulb-an upright series of fleshy overlapping leaf bases attached to a small basal stem, as in the onion

corm-an upright, hard or fleshy stem surrounded by dry scaly leaves, as in the gladiolus "bulb"

rhizome-a horizontal stem with reduced scaly leaves, as in many grasses

tuber-an enlarged fleshy tip of an underground stem, as in the Irish potato

turion-a swollen scaly offshoot of a rhizome

TERMINOLOGY APPLIED TO LEAVES

In most vascular plants, the leaf is a broad flattened photosynthetic organ. A wide variety of taxonomic features is derived from it. The point at which a leaf is inserted on the stem is the **node**. The region between two successive nodes is the **internode**.

PARTS OF THE LEAF

blade (=**lamina**)-the flattened expanded portion; a few leaves are bladeless

petiole-the stalk which supports the lamina; if missing, the leaf is **sessile**

stipules-a pair of appendages located at the base of the petiole where it joins the stem; often short-lived and seen only as stipule scars; if not formed, the leaf is **exstipulate.**

ARRANGEMENT (=PHYLLOTAXY)

alternate-one leaf at a node

basal-at the base of the plant, the internodes being much reduced

cauline-the leaves of a well-developed stem, as opposed to basal leaves

decussate-opposite leaves which alternate at right angles to one another at successive nodes, thereby forming 4 rows of leaves

distichous-two-ranked on opposing sides of the stem and in the same plane

equitant-leaves folded about one another in two ranks, as in many members of the Iridaceae

fascicled-clustered

imbricate-overlapping

opposite-two leaves at a node

rank-a vertical row of leaves (or flowers)

rosette-a radiating leaf cluster at or near the base of the plant

rosulate-in rosettes

whorled (=**verticillate**)-three or more leaves at a node

COMPOSITION

The term **composition** is used to denote whether the leaf blade is represented by a single segment (**simple**) or is divided into two or more discrete segments (**compound**). Simple leaves may exhibit various degrees of indentation, thereby approaching compound leaves.

lobed-indented about one-fourth to almost half way to the midrib or base of the blade

cleft-indented about half way to the midrib or base of the blade

parted-indented nearly all the way to the midrib or the base of the blade

divided-indented to the midrib or base of the blade

Compound leaves are divided into separate segments called **leaflets**. The stalk of a leaflet is a **petiolule**. If the compound leaf has a **central** axis, it is referred to as a **rachis**. Several orders of compounding are recognized.

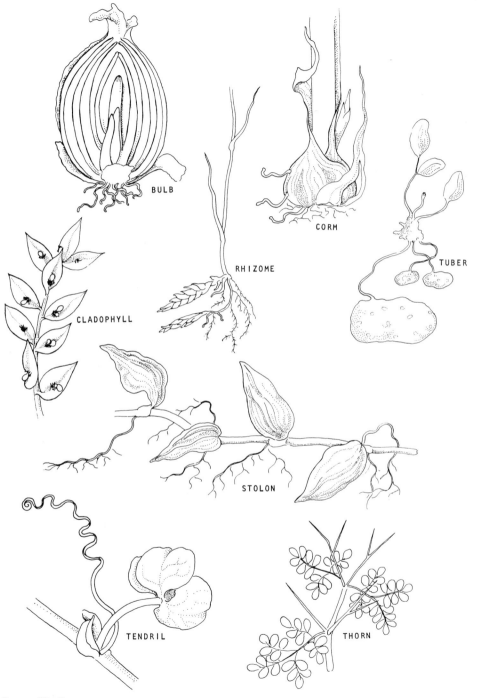

Fig. 7. **Stem modifications.**

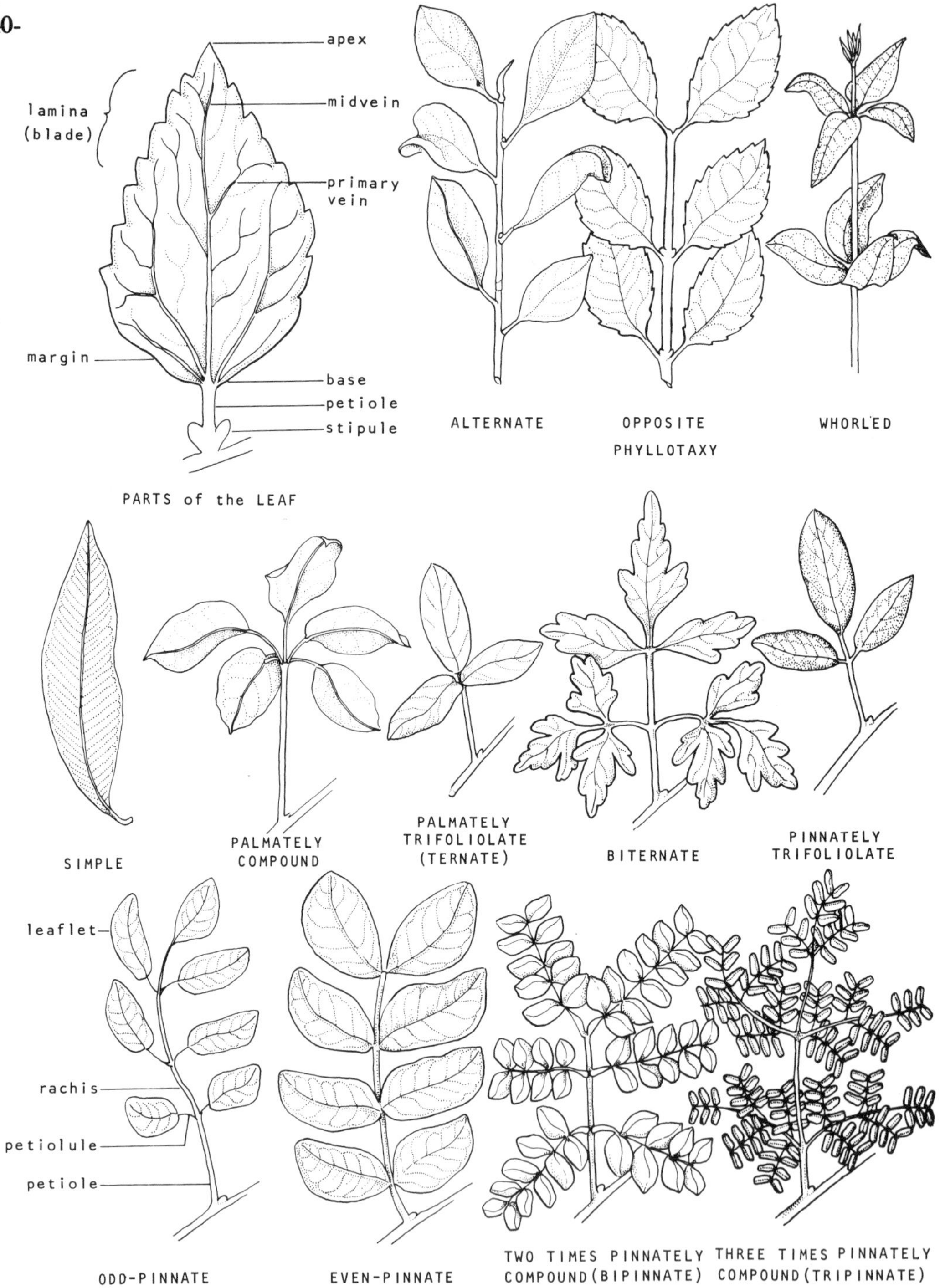

apex

midvein

lamina (blade)

primary vein

margin

base
petiole
stipule

PARTS of the LEAF

ALTERNATE   OPPOSITE   WHORLED

PHYLLOTAXY

SIMPLE

PALMATELY COMPOUND

PALMATELY TRIFOLIOLATE (TERNATE)

BITERNATE

PINNATELY TRIFOLIOLATE

leaflet

rachis

petiolule

petiole

ODD-PINNATE

EVEN-PINNATE

TWO TIMES PINNATELY COMPOUND (BIPINNATE)

THREE TIMES PINNATELY COMPOUND (TRIPINNATE)

Fig. 8. **Leaf parts and composition.**

once-pinnately compound-the blade divided only once into first order leaflets, these arranged along a rachis; if a terminal leaflet is present, the leaf is **odd-pinnate,** if it is absent the leaf is **even-pinnate.**

twice-pinnately compound (=bipinnate)-the blade divided twice so that the first order leaflets are themselves further divided into second order leaflets

thrice-pinnately compound (=tripinnate)-the blade divided three times so that first-, second-, and third- order leaflets are present

decompound-a compound leaf that is at least twice-pinnately compound; used particularly when the blade is divided into numerous segments, as in some members of the Umbelliferae

palmately (=digitately) compound-the blade divided into leaflets which radiate from the apex of the petiole, no rachis present

ternate-a palmately compound leaf with three leaflets

biternate-a ternate leaf in which the first order leaflets are themselves ternately compound

trifoliolate-a compound leaf with three leaflets, these either pinnately or palmately disposed depending upon the relative lengths of the petiolules; not to be confused with **trifoliate,** a plant with three leaves, as in *Trillium.*

OVERALL SHAPE

Many generalized leaf shapes have been given names. While fine examples of all of these may be found, many intermediates will also be encountered. In such instances, merely hyphenate the two shapes, an in linear-oblong.

acicular-needle-shaped

cordate-of the shape of the stylized heart, with the petiole attached between the basal lobes

deltoid-of the shape of an equilateral triangle

elliptic-oval, the ends rounded and widest at the middle

falcate-sickle-shaped

filiform-thread-like

hastate-more or less arrowhead-shaped, but with the basal lobes divergent

lanceolate-lance-shaped, several times longer than wide; the sides curved, with the blade broadest below the middle

linear-several times longer than wide, the sides more or less parallel

lyrate-with a series of pinnate lobes and a larger terminal lobe

obcordate-as in the cordate leaf, but the petiole attached at the point of the heart

oblanceolate-as in the lanceolate leaf, but the petiole attached at the narrow end

oblong-about two or three times longer than broad; rectangular with rounded corners

obovate-as in ovate, but the petiole attached at the narrow end

orbicular (=rotund)-circular or nearly so

oval-broadly elliptic, the length less than twice the width

ovate-the shape of the longitudinal section through a chicken's egg, with the petiole attached at the broad end

reniform-kidney-shaped or bean-shaped

runcinate-coarsely-toothed, the teeth pointing toward the base of the leaf, as in the dandelion

sagittate-arrowhead-shaped

spatulate (=spathulate)-spoon-shaped

subulate-slender and tapering to a point, as in the awl, a tool used to make holes in leather

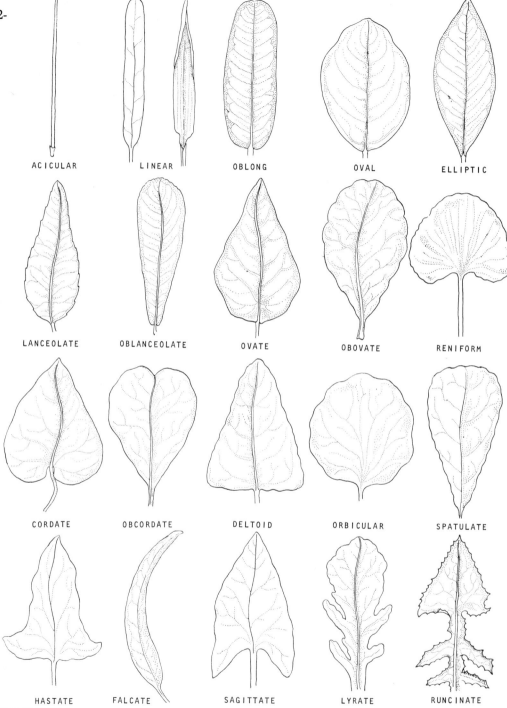

ACICULAR   LINEAR   OBLONG   OVAL   ELLIPTIC

LANCEOLATE   OBLANCEOLATE   OVATE   OBOVATE   RENIFORM

CORDATE   OBCORDATE   DELTOID   ORBICULAR   SPATULATE

HASTATE   FALCATE   SAGITTATE   LYRATE   RUNCINATE

Fig. 9. **Leaf shapes.**

LEAF MARGINS

As with the generalized leaf shapes, margins may closely approximate the types defined below or may be somewhat intermediate.

ciliate-with fine hairs on the margin

crenate-scalloped, the teeth blunt

dentate-with coarse angular teeth directed outward at right angles to the margin

denticulate-finely dentate

doubly serrate (=**biserrate**)-with the serrations themselves serrate

entire-not in any way indented, the margin featureless

erose-gnawed, as if chewed upon

fimbriate-fringed, the hairs coarser than in ciliate

incised-deeply and sharply cut

laciniate-slashed into narrow pointed segments

revolute-the margin rolled toward the lower side of the blade

serrate-with coarse saw-like teeth which point forward

serrulate-finely serrate

sinuate-wavy in and out, in the plane of the blade

undulate (=**crisped**)-wavy perpendicular to the plane of the blade

VENATION

The venation pattern is determined by the pattern of the principal veins of the leaf blade. The following are commonly recognized:

parallel-several to many veins of about the same size (the midrib sometimes more conspicuous) and parallel to one another, as in many monocots

pinnate-prominent midvein with a series of major veins arising at about 30-45^0 angles along its length

net-a complex venation pattern of major and minor veins which form a network or reticulum

palmate-the major veins radiating from a common point at the base of the blade, as in the maples

arcuate-an uncommon pattern in which the major veins curve gently upward

LEAF APICES

The leaf apex may closely approximate the types defined below or may be somewhat intermediate.

acute-apex formed by two straight margins meeting at less than 90^0

acuminate-an acute apex whose sides are concave and taper to an extended point

apiculate-terminating in a short, sharp, flexible point

aristate-an abrupt, hard, bristle-like point (**awn**)

attenuate-the apex drawn out into a long gradual taper

caudate-the apex tail-like

cuspidate-a sharp-pointed tip formed by abruptly and sharply concave sides

emarginate-with a shallow notch at the apex

mucronate-with a hard, short, abrupt point (**mucro**)

obtuse-apex formed by two lines which meet at more than a right angle

rounded-the apex gently curved

spinose-with a spine at the tip

truncate-the apex appearing chopped off

LEAF BASES

Several of the same terms used to define leaf shapes and apices are also used to define leaf bases. These include **acute, attenuate, cordate, hastate, obtuse, rounded, sagittate,** and **truncate.** Additional types are presented below.

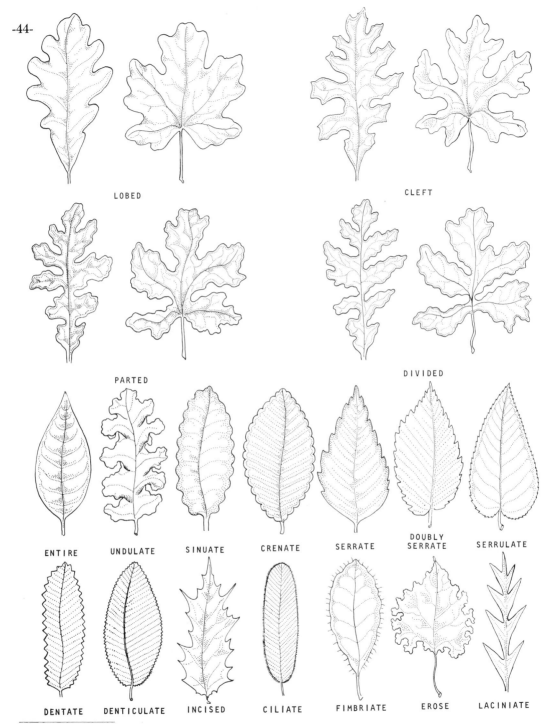

Fig. 10. **Leaf margins.**

auriculate-with a pair of rounded lobes which somewhat resemble the human ear

clasping-the bases partly to completely surrounding the stem

cuneate-wedge-shaped

oblique-asymmetrical; unequal-sided

perfoliate-the condition of a sessile leaf when the base completely encircles the stem

LEAF MODIFICATIONS

The following leaf modifications are commonly recognized.

bract-a leaf much reduced in size, particularly if it is associated with a flower or inflorescence

phyllode (=phyllodium)-a leaflike petiole of a bladeless leaf, as in some *Acacia*

sheath-the basal portion of the leaf which surrounds the stem, as in members of the Commelinaceae

spine-a leaf or portion of a leaf which is sharp-pointed, the most common examples being paired stipular spines; not to be confused with **thorns,** which are modified stems or **prickles,** which are mere outgrowths of the epidermis, as in the cultivated rose.

tendril-a twining leaf or portion of a leaf, as in the leaflets of the sweet pea; tendrils may also be of stem origin

SURFACES AND SURFACE COVERINGS

The surface of stems, leaves, flowers, fruits and seeds reveals many useful features. The following terms are applied:

alveolate-resembling the surface of a honeycomb

dull-not shining, lacking lustre

glabrate-hairy at first, but then glabrous

glabrous-without hairs

glaucous-covered with a whitish waxy bloom

lustrous-shining

papillate (=papillose)-with pimple-like protuberances

pitted-covered with small cavities

punctate-dotted with pin-point impressions or translucent dots

reticulate-netted with regular slightly elevated lines

rugose-wrinkled

scurfy-covered with minute scales

striated-marked with longitudinal lines

sulcate-furrowed with longitudinal channels

tuberculate (=verrucose)-warty

viscid-sticky

These surfaces, no matter what their type, may be clothed with hairs of some sort. Such surface coverings are technically referred to as **vestiture** or **vesture.** The types of vestiture are numerous and perhaps more subjectively defined than the other features described in this chapter. I have grouped vestiture types in an attempt to facilitate distinguishing them. Once you have mastered this subject you are ready to turn to the theological question, "How many angels can dance on the head of a pin"?

A. Hairs simple, unbranched
  B. Hairs hooked or barbed

      barbellate- hairs with barbs down the sides

      glochidiate- hairs barbed at the tip only, as in the hairs of certain cacti

     uncinate- hooked hairs

  B. Hairs without hooks or barbs

    C. Hairs restricted to an apex or margin, not superficial

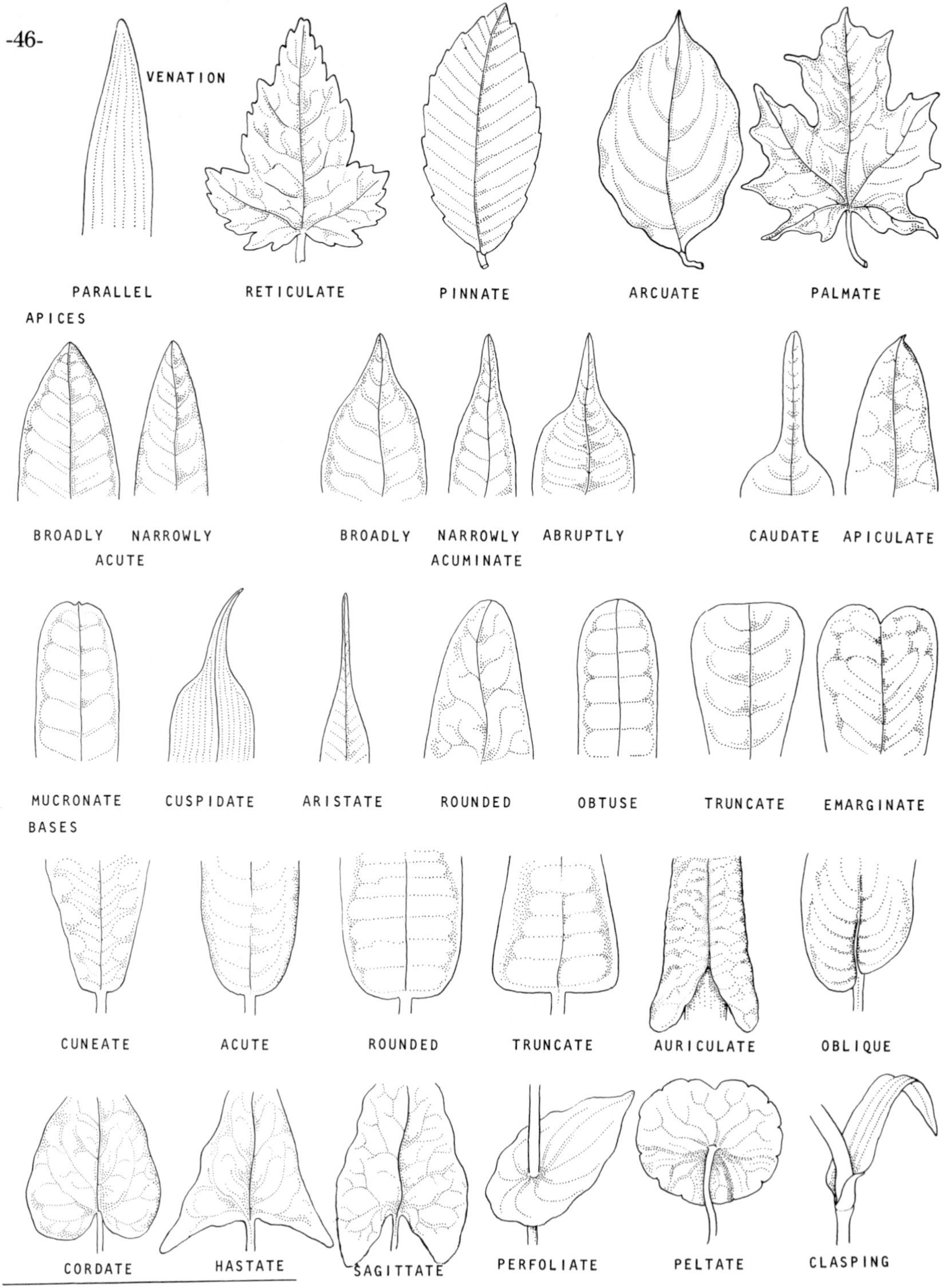

VENATION

PARALLEL   RETICULATE   PINNATE   ARCUATE   PALMATE

APICES

BROADLY   NARROWLY   BROADLY   NARROWLY   ABRUPTLY   CAUDATE   APICULATE
        ACUTE   ACUMINATE

MUCRONATE   CUSPIDATE   ARISTATE   ROUNDED   OBTUSE   TRUNCATE   EMARGINATE

BASES

CUNEATE   ACUTE   ROUNDED   TRUNCATE   AURICULATE   OBLIQUE

CORDATE   HASTATE   SAGITTATE   PERFOLIATE   PELTATE   CLASPING

Fig. 11. **Leaf venation, apices, and bases.**

ciliate- hairs along the margins only

comose- with a tuft of hairs at the apex of a seed or at the base of a floret in a grass spikelet

fimbriate-as in ciliate, but coarser and longer

C. Hairs scattered over the surface, not restricted to margins

D. Hairs curled, interwoven or entangled

arachnoid- slender, white, loosely tangled hairs; cobwebby

floccose- with tufts of soft hairs which rub off easily

lanate- woolly or cottony

tomentose- densely and softly matted

D. Hairs short and squat to long and delicate, but not curled and interwoven

canescent (= **hoary**)- with a dense mat of greyish-white hairs

echinate- with straight, often comparatively large prickle-like hairs

glandular- hairs with swollen tips; gland-bearing

hirsute- with rough or coarse, more or less erect hairs

hirtellous- minutely hirsute

hispid- with long, rigid, bristly hairs

mealy (= **farinaceous** or **farinose**)- swollen hairs which collectively form a covering resembling cooking meal

papillate (= **papillose**)- hairs pimple-like

pilose-with sparse, slender, soft hairs

puberulent- minutely canescent

pubescent- downy; the hairs short, soft, and erect

scabrous- rough to the touch because of coarse, stiff, ascending hairs

sericeous- silky; the hairs long, fine, and appressed

setaceous (= **setose**)- bristly

strigose- hairs sharp, appressed, rigid, and often swollen at the base

velutinous- velvety; the hairs dense, firm, and straight

villous- shaggy; the hairs long, slender, soft, but not matted

A. Hairs branched or forked

malpighiaceous (= **dolabriform**)- forked hairs attached at the middle, as in a miner's pick

stellate- with a few to several branches or "arms" which arise from a central point; may be sessile or stalked

PAPILLATE PITTED SCURFY

UNCINATE BARBELLATE GLOCHIDIATE

VELUTINOUS TOMENTOSE LANATE FLOCCOSE

SCABROUS STRIGOSE GLANDULAR FARINOSE OR MEALY

HIRSUTE HIRTELLOUS HISPID ECHINATE

PUBERULENT PUBESCENT PILOSE VILLOUS

SERICEOUS DOLABRIFORM/MALPIGHIACEOUS STELLATE

Fig. 12. **Surfaces and Vestitures.**

5- Flowers, Inflorescences, and Fruits

If most of us were asked to state the character that sets the flowering plants apart from other vascular plants, we would quickly answer that it is "the flower". The reply seems so obvious that we might wonder why the question was even asked. How disillusioning it is to find that so straightforward an answer is relatively meaningless. The flower, the basic reproductive unit of the flowering plants, cannot be morphologically defined in such a way that similar reproductive structures in the gymnosperms are excluded. By the way, the answer to the question is that there is no single feature that separates the flowering plants from other groups. None is both unique and universal within the angiosperms. Endosperm development and the sieve tube-companion cell relationship probably have higher predictive value because of the variation in flower structure.

From a morphological standpoint, a typical flower is a stem tip which bears two series of appendages that are sterile and two that are fertile. Following the generally accepted view, all four series are considered modified leaves. The stem tip upon which the four series are borne is the **receptacle** or **torus.** In more primitive flowering plants the receptacle is elongate or domeshaped. In more advanced forms it is gently rounded to flat. The sterile appendages are typically of two types, the **sepals** and **petals.** The sepals, the lowermost and outermost appendages, constitute a series called the **calyx.** The petals, the inner set of sterile structures, contitute the **corolla.** The calyx and corolla together make up the **perianth.** The fertile appendages are also typically of two types, the **stamens** and the **carpels.** They are homologous with the **microsporophylls** and **megasporophylls,** respectively. The stamens, situated just inside the perianth, make up the **androecium.** The carpels, innermost and uppermost on the receptacle, constitute the **gynoecium.**

Many flowers produce nectar, a sweet fluid attractive to animal pollinators. The structures within the flower which produce nectar are called **nectaries.** They may be modified portions of sepals, petals, stamens, carpels, or receptacle.

PRESENCE OR ABSENCE OF FLORAL SERIES

Terms applied to individual flowers:

complete-a flower which has all four series

incomplete-a flower with one or more series missing

perfect (=**bisexual**)-a flower with both stamens and carpels, without regard to the state of the perianth

imperfect (=**unisexual**)-a flower with either stamens or carpels, but not both, without regard to the state of the perianth

staminate (=**male**)-a unisexual flower in which stamens are present, the carpels being rudimentary or suppressed

pistillate (=**female**)-a unisexual flower in which only carpels are present, the stamens being rudimentary or suppressed

Terms applied to plants with imperfect flowers:

monoecious-a species in which any plant bears both staminate and pistillate flowers

dioecious-a species in which any particular plant bears either staminate or pistillate flowers, but not both; the species is composed of separate staminate and pistillate plants

polygamous-a species in which the plants bear both perfect and imperfect flowers; **polygamo-monoecious** and **polygamo-dioecious** conditions may be recognized

NUMERICAL BASIS OF FLOWER PARTS

The number of sepals, petals, stamens, and carpels is typically rather uniform within a particular species. Most monocot flowers have these series in 3's or multiples thereof, while most dicots have the floral parts in 4's or 5's or multiples thereof. If a flower is based upon 3's, it is said to be **3-merous;** if on 4's, then **4-merous,** etc.

RELATIVE INSERTION OF FLORAL PARTS

If we look down upon an open-faced flower, we will see that the point of insertion of the sepals alternates with that of the petals, that the petals alternate with the insertion of the stamens, and that the stamens alternate with the carpels. This alternation of floral series can be very useful when determining the identity of a missing set of flower parts.

The vertical insertion of the gynoecium in relation to the other floral series is the basis of terminology used in all keys and descriptions. In the **hypogynous** flower, the sepals, petals, and stamens are inserted beneath the ovary. Because the ovary is above the point of insertion on the other three series, it is said to be **superior.**

In a **perigynous** flower, the sepals, petals, and stamens are united by their lower portions to form an open cup-like structure called a **hypanthium.** The gynoecium sits inside the hypanthium and is not fused to it in any way. Because the hypanthium is formed from perianth and androecium which are eventually inserted below the gynoecium, the ovary is said to be superior, as in the hypogynous flower.

In the **epigynous** flower, the sepals, petals, and stamens seem to arise from a point above the insertion of the ovary, so that it is **inferior** to them. A hypanthium may be completely fused around the ovary, as in the apple or pear, or it may extend above the ovary as a free hypanthium, as in the fuchsia. In other instances, it appears that the gynoecium has partially sunken into the receptacle. The ovary may be termed **half-inferior** if a hypanthium or receptacle extends only half way up the ovary.

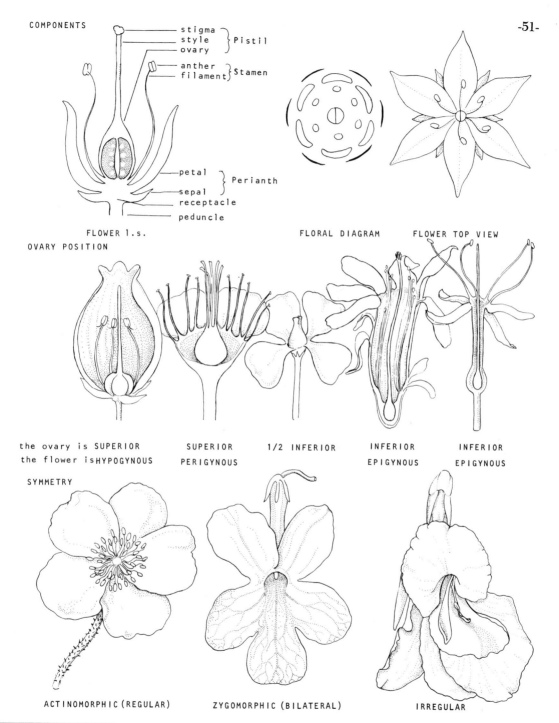

COMPONENTS

stigma ⎫
style ⎬ Pistil
ovary ⎭

anther ⎫ Stamen
filament ⎬

petal ⎫
⎬ Perianth
sepal ⎭
receptacle
peduncle

FLOWER l.s.

FLORAL DIAGRAM

FLOWER TOP VIEW

OVARY POSITION

the ovary is SUPERIOR
the flower is HYPOGYNOUS

SUPERIOR
PERIGYNOUS

1/2 INFERIOR

INFERIOR
EPIGYNOUS

INFERIOR
EPIGYNOUS

SYMMETRY

ACTINOMORPHIC (REGULAR)

ZYGOMORPHIC (BILATERAL)

IRREGULAR

Fig. 13. **Floral morphology.**

THE PERIANTH

The perianth is composed of all of the sepals and petals. In most flowers, the sepals are green and the petals are white or some color other than green. In those instances where sepals and petals are the same color, the two series may be distinguished by their points of insertion on the receptacle. The outer ones are, by definition, sepals. In other instances it is not convenient to distinguish sepals and petals, becuase they are inseparable by size, shape, color, or texture. Here the term **tepal** may be applied, particularly when there are numerous perianth segments.

Miscellaneous terms applied to the perianth:

anterior lobe (=**abaxial lobe**)-the lobes away from the inflorescence axis

beard-a tuft or line of hairs on a perianth segment, as in *Iris*

bristle-the much reduced hair-like perianth in some sedges

calcarate-spurred

callosity-a thickened raised area on a perianth segment, as in *Rumex*

carinate-keeled, as in many legume flowers

claw-a long narrow stalk-like base of a petal or sepal

clawed-having a claw

corniculate-bearing a small horn-like protuberance, as in the milkweed flower

corona-any outgrowth situated between the corolla and the androecium, as in the milkweeds or narcissus

cruciform-cross-shaped, as in the sepals and petals of the mustard family

galea-a helmet, as in the upper sepal of the monk's hood

galeate-having a galea

gibbosity-the inflated portion on one side of a calyx or corolla near its base, as in the snapdragon

gibbous-having a gibbosity

hood-a component of the corona in the milkweed flower

keel-a structure resembling the bottom of a boat, as in the two fused lower petals of many legume flowers

keeled-having a raised ridge, as in the keel of a boat

limb-the upper expanded portion of the calyx or corolla situated above the tube, throat or claw

lip (=**labellum**)-the lower, often highly modified petal of an orchid flower; the upper or lower portion of a calyx or corolla which is bilaterally symmetrical, as in many mints

lobe-a part of a perianth segment that represents a major division

lodicule-the highly reduced perianth of the grasses

palate-a raised area or rounded projection which more or less obstructs the throat of a sympetalous corolla

pappus-the reduced perianth of the aster family (Compositae)

petaloid-petal-like

posterior lobe (=**adaxial lobe**)-the lobe next to the inflorescence axis

saccate-pouch-like, as in the lower petal of some orchids

sepaloid-sepal-like

spur-a sac-like or tubular projection from the perianth, as in the columbines; often associated with nectaries

standard (=**banner**)-the upper petal of a papilionoid or of a caesalpinioid legume

ventricose-inflated on one side of the calyx or corolla at about the middle; see also **gibbous**

Presence or absence of perianth parts:

apetalous-without petals

asepalous-without sepals

naked (=**achlamydeous**)-without sepals or petals

biseriate (=**dichlamydeous**)-with a perianth of both sepals and petals

uniseriate (=**homochlamydeous**)-with a perianth of a single series, the components arbitrarily designated as sepals or tepals

Fusion of perianth parts:

aposepalous (=**chorisepalous**)-with separate sepals

apopetalous (=**choripetalous** = **polypetalous**)-with separate petals

distinct- separateness of like parts, as in the petals from one another or the sepals from one another

synsepalous-with fused sepals

sympetalous (=**gamopetalous**)-with fused petals

connate-fusion of like parts, as in petals fused to one another

calyx tube-the tube of a synsepalous calyx

corolla tube-the tube of a sympetalous corolla

As you might expect, there are intermediate situations in which sepals and petals are neither completely fused, nor are they completely separate. Most authors will apply the terms synsepalous and sympetalous if the parts are connate through much or most of their length, even if their tips or lobes are distinct. The same terms that were applied to leaves to describe degrees of division, **lobed, cleft, parted,** and **divided,** may also be used for sepals and petals.

Symmetry:

actinomorphic (=**radial** = **regular**)-sepals and petals radiating from the center of the flower; the sepals are all similar to one another, as are the petals to one another

zygomorphic (=**bilateral**)-the perianth constituted in such a way that only a median plane will yield two equal halves, as in most orchids or the sweet pea flower

irregular-the perianth constituted in such a way that it is not possible to divide it into two equal halves, as in the canna flower

Corolla Configurations:

The corolla of many flowers may be differentiated into a basal cylindrical **tube,** a **throat** which is the opening at the top of the tube, and a **limb** which is the expanded uppermost portion of the corolla. The relative proportions of these parts coupled with floral symmetry yield corolla shapes or configurations which bear names.

Actinomorphic configurations:

campanulate-bell-shaped

cruciate-cross-shaped, as in most members of the mustard family

funnelform (=**infundibular**)-the elongate tube gradually widening into the limb, as in the morning glory family

rotate-saucer- or wheel-shaped; the tube very short, the throat and limb abruptly flared

salverform-the tube elongate, the throat and limb abruptly flared

subglobose-more or less spherical

tubular-throat and limb only slightly wider than the elongate tube

urceolate-urn-shaped, as in many members of the heath family

Zygomorphic configurations:

bilabiate-literally "two-lipped", with some petals forming an upper lip, while the others form a lower lip, as in many mints

-54-

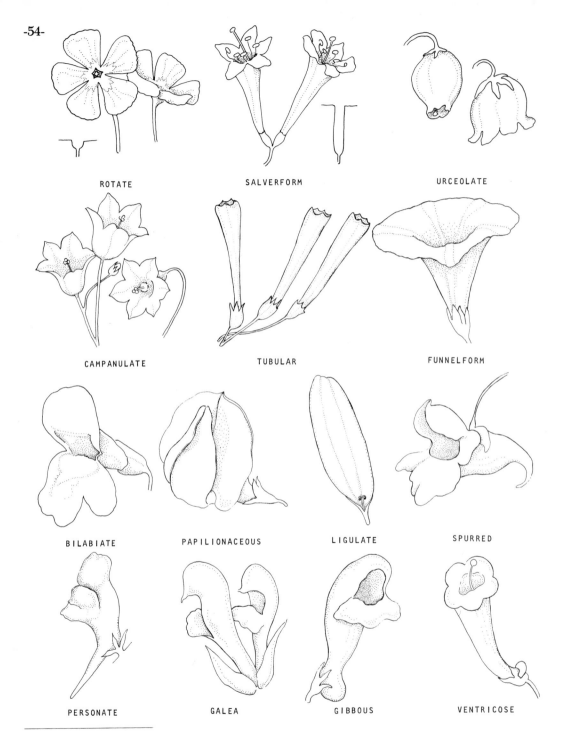

Fig. 14. **Corolla configurations.**

calceolate-slipper-shaped, as in the lady-slipper orchids

ligulate-strap-shaped, as in the corolla of the ligulate flowers of many members of the aster family

papilionaceous-the flower of many members of the pea family, with a broad upper petal (the standard), two lateral petals (the wings), and two lower petals (the keel petals)

personate-a bilabiate flower in which the upper lip is arched and the lower protrudes into the throat of the corolla, as in the toadflax

Please note that spurs, gibbosites, and ventricosites will also produce a zygomorphic condition.

THE ANDROECIUM

The androecium is composed of one or more stamens. All of the stamens of a flower may be well-developed and fertile or one to several of them may be rudimentary. Sterile, often antherless stamens, are called **staminodes** or **staminodia.** They may resemble vestigial stamens or they may be more heavily modified and resemble perianth segments.

The stamen in most flowering plants is differentiated into a narrow **filament** or stalk and a broader sporogenous **anther.** The anther is composed of two **anther-sacs** or **thecae,** separated by the **connective,** an extension of the filament between the two components. The connective may be inconspicuous to quite broad and easily seen. The following specialized terminology is applied to the androecium:

Attachment of the anthers:

basifixed-one end of the anther appearing terminal on the filament

dorsifixed-anther attached on its side to the filament

versatile-anther attached as in dorsifixed, but free-moving on the filament

Fusion of stamens to one another:

monadelphous-stamens united by their filaments into a single group

diadelphous-stamens united by their filaments into two groups

polydelphous-stamens united by their filaments into several groups

column-the structure formed by fused filaments

syngenesious-stamens united by their anthers

Relative stamen length:

didynamous-an androecium of four stamens, with two of them on longer filaments, as in the mint and figwort families

tetradynamous-an androecium of six stamens, four of them on filaments longer than the other two, as in the mustard family

Miscellaneous terms:

appendaged-modifications of the anther apex, base or filament into horns, tips, tails or other protuberances

caudate-with tail-like appendages

exserted-the stamens extending from the throat of the corolla

included-the stamens enclosed in the throat of the corolla

pollinium-a pollen mass, as in the milkweeds and orchids

staminal disk-a raised fleshy ring or cushion formed by the fusion of staminodes or nectaries on which the stamens sit

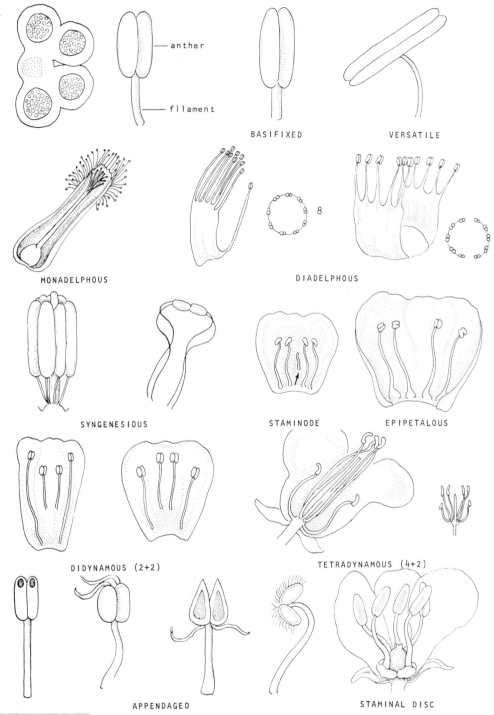

Fig. 15. **The androecium.**

THE GYNOECIUM

The gynoecium is composed of carpels, the ovule-bearing parts of the flower. If we adopt the most widely accepted theory, a carpel is an enrolled seed-bearing leaf, a megasporophyll. We may recognize three basic types of gynoecia. The **unicarpellate** gynoecium is composed entirely of a single carpel, as in the pea family. The **apocarpous** gynoecium is composed of two or more carpels, each separately inserted on the receptacle. The carpel number may be quite high, as in the buttercup family. The only remaining possibility is two or more carpels fused together to form a **syncarpous** gynoecium. The number of fused carpels is often two, three or four, as in the figwort, lily, and evening primrose families, respectively. In a few syncarpous gynoecia, the number of carpels may be somewhat higher, as in the poppy family. It should be noted that the carpels may be so completely fused to one another that there is little evidence of syncarpy.

A typical carpel is differentiated into three parts. The upper portion is the **stigma,** the area receptive to pollen. It may be variously lobed, divided, flattened or rounded. Below the stigma is the columnar or neck-shaped **style.** In a few flowers the style is not well-developed. At the base of the carpel is the swollen seed-bearing portion, the **ovary.** It is usually globose or cylindrical. Within the ovary is the region along the inside wall, the **placenta,** where the seeds or ovules are attached. For reasons which will become clear later, it is useful to think of a carpel in terms of a single stigma, a single style, and a single placenta.

Many authors have chosen to present the seed-bearing part of the flower in terms of a basic unit called the **pistil.** According to this terminology, a flower may have a **simple pistil,** or a **compound pistil** made up of two or more united pistils. I find the "pistil terminology" confusing and unnecessary. It is confusing because the term pistil is ambiguous. Sometimes it is synonymous with carpel; at other times, gynoecium. The two systems are compared in Table 4 below.

Placentation types:

If we view the ovary in cross-section, it may exhibit the following parts:

locule-the chamber within the ovary; there may be one or more locules

ovule-immature seed

placenta-the region or line along which the ovules are attached

septum-an interior wall which separates the locules in those instances where two or more chambers occur

In the unicarpellate gynoecium, there is a single locule and only one placenta. It is located along the **ventral suture,** the line marking the union of the two edges of the enrolled seed-bearing leaf, the carpel. The position of the placenta or **placentation type** is said to be **marginal.** The same type is found in each of the separate carpels of an apocarpous gynoecium.

The situation in the syncarpous gynoecium is more complex. In many flowers, the ovary has a single locule, but there are two or more

Table 4. Comparison of carpel and pistil terminology

Number of Carpels	Carpel terminology	Pistil terminology
1	unicarpellate	simple pistil
2 or more, separate	apocarpous	each a pistil
2 or more, fused	syncarpous	compound pistil

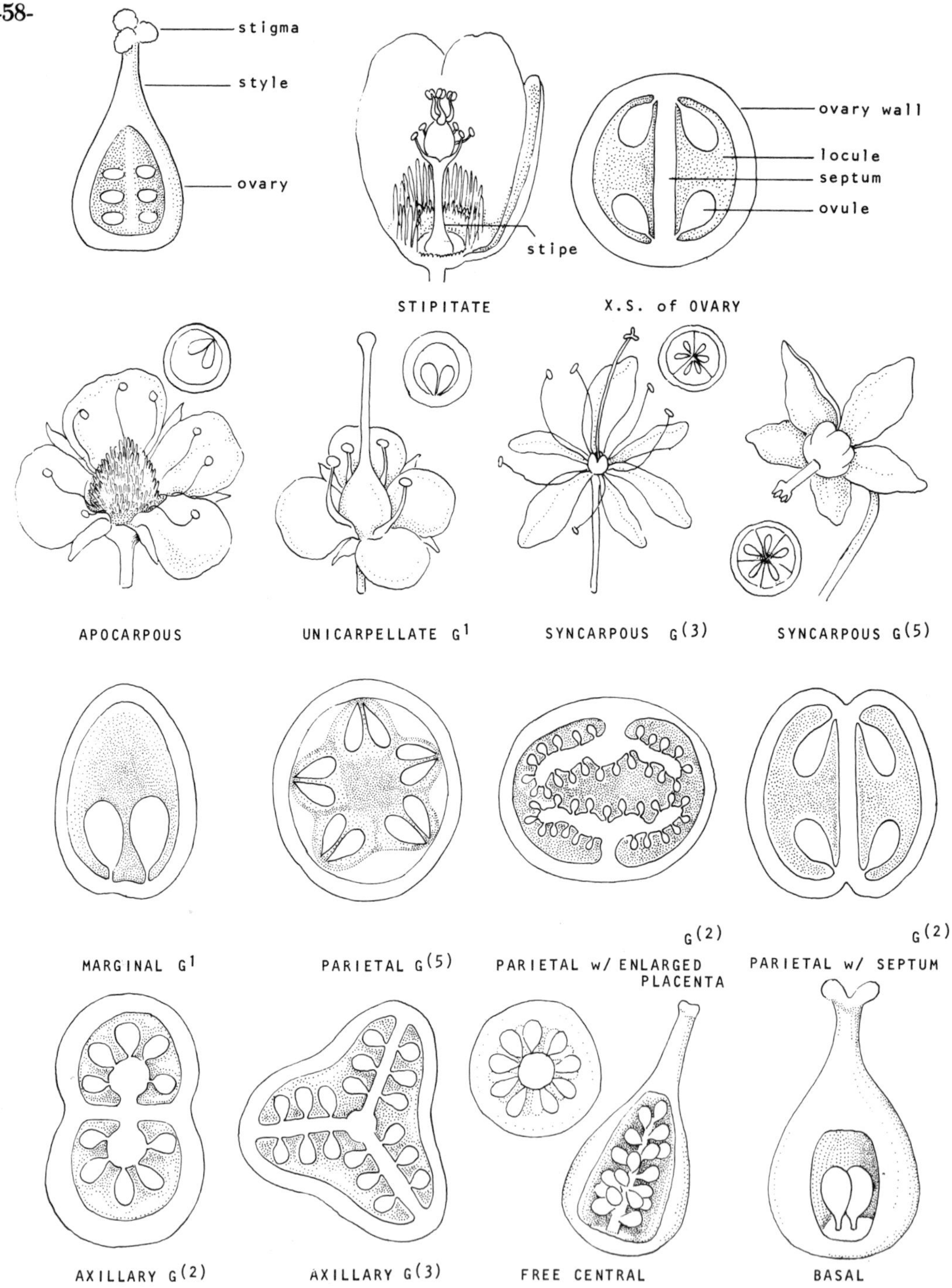

Fig. 16. **The gynoecium.**

placentae arranged along the inside of the wall. In such a situation, the placentation type is **parietal.** Many other flowers have ovaries with internal walls which separate them into two or more locules. The placentae appear at the center of the ovary, either at the midpoint of a single septum or where several septa fuse, if there is more than one. This pattern is **axile** or **axillary** placentation. A more unusual situation is to find an ovary with no interior walls, but a central column which is more or less covered by ovules. The column is attached to the ovary wall only at its base. The placentation type is **free-central.** The final situation is seen in the ovary with a single locule and a greatly reduced number of ovules, often just one, sitting at the base of the chamber. This is **basal** placentation.

Determination of carpel number:

It is relatively easy to determine the number of sepals, petals, and stamens. The fusion of these parts is seldom so complete that great uncertainty arises. This is not the case in determining the number of carpels which compose the gynoecium. Keep in mind that a carpel may be defined in terms of a single stigma, style, and placenta. External evidence of carpel number may be seen in the number of stigma lobes, the number of styles or lobes of the ovary. However, in may syncarpous gynoecia the stigma is unlobed, the styles completely fused to one another, and no lobing of the ovary is evident. In situations of this kind, it is necessary to make a cross-section of the ovary to be certain. If the placentation type is parietal, basal, or free - central the single locule will not be indicative of carpel number. You must count the number of placentae. If the placentation is axillary, the number of locules may be indicative, but it is still best to count placentae.

FUSION OF TWO OR MORE FLORAL SERIES

In the same way that sepals may be fused to one another or stamens may be connate by their filaments, so may entire series be joined to one another. The following terms are applicable:

adherent- superficially joined, but actually only touching

adnate-organically fused

coalesced-partially fused in a somewhat irregular manner

free-separate, not joined

column (=**gynandrium** = **gynostemium**)-the structure resulting from the adnation of stamens and gynoecium in the orchids

epipetalous-arising from the corolla, as in stamens inserted within the corolla tube

gynostegium-the structure resulting from the adnation of the stamens and gynoecium in the milkweeds

hypanthium (=**floral cup**)-the cup-like structure resulting from the fusion of the lower portions of the sepals, petals, and stamens; it is considered that each of these series is separately inserted on the rim of the hypanthium, so that synsepaly, sympetaly, and monadelphy are not present.

FLORAL FORMULAS

It is possible to display a great deal of information about the structure of a flower by using a system of notation called the **floral formula.** The conventions used in writing them have not been standardized, so that references will vary in certain details. The idea is a very simple one. Let the four floral series (calyx, corolla, androecium, and gynoecium) be represented by K, C, A, and G. Exponents indicate the number of parts. Many refinements are possible. The examples below illustrate how a floral formula may be constructed.

Components:

K= calyx

C= corolla

A = androecium

G = gynoecium

Number of parts:

K^5 = calyx of 5 sepals

A^{10} = androecium of 10 stamens

A^{4+2} = androecium of 6 stamens, 4 in one set, 2 in another

$G^{2\text{-}3}$ = gynoecium of 2 or 3 carpels

A^x = androecium of a low unstable number (ca. 10-20) stamens

G^∞ = gynoecium of a high unstable number; well over 20 carpels

C^0 = corolla missing; petals 0

$A^{4[5]}$ = androecium usually of 4 stamens, rarely 5

Symmetry:

K = calyx actinomorphic

C = corolla actinomorphic

Kz = calyx zygomorphic

Cz = corolla zygomorphic

Fusion of parts:

K^4 = calyx of 4 separate sepals

C^3 = corolla of 3 separate petals

$K^{(4)}$ = calyx of 4 fused sepals

$G^{(2)}$ = gynoecium of 2 fused carpels

G^4 = gynoecium of 4 separate carpels

$A^{\overset{\frown}{5}}$ = androecium of 5 stamens fused by their filaments

$\overset{\frown}{A^5 \ G^2}$ = androecium and gynoecium fused by their upper parts

$\underline{K\ C\ A}$ = calyx, corolla, and androecium united into a hypanthium

Ovary position:

$\underline{G}$ = ovary superior; flower hypogynous or perigynous

$\overline{G}$ = ovary inferior; flower epigynous

$\text{-}G\text{-}$ = ovary half-inferior

$\overline{\underline{G}}$ = ovary either superior or inferior

Thus the floral formula

$$K^5 \ Cz^{1+(2)\ +2} \ A^{\underline{9}\ +1} \ \underline{G}^{\ 1}$$

may be translated: calyx of 5 separate sepals; corolla zygomorphic, of 5 petals, 2 joined together, 2 free and forming a pair, and a 5th different from the others; androecium of 10 stamens, 9 joined by their filaments, the 10th free; gynoecium unicarpellate, the ovary superior.

INFLORESCENCE TYPES

An **inflorescence** is an arrangement of one or more flowers on a floral axis. The number of flowers, their positional relationships, the degree of development of their pedicels, and the nature of the branching pattern within the flower cluster determine the particular inflorescence type. The following general terms are applied to parts of inforescences:

axillary- arising from the axil of a leaf

bract-a reduced leaf, particularly one associated with a flower or subtending the inflorescence

bracteole-a small second-order bract

cupule-fused involucral bracts subtending a flower, as in an acorn cup

epicalyx-a set of sepaloid bracts

involvucel -a secondary series of bracts, often subtending a subunit of an inflorescence

involucre-an organized set of bracts, either separate from one another or fused to form a cup

pedicel-the supporting stalk of an individual flower in a multi-flowered inflorescence; **pedicellate** -having a pedicel

peduncle-the supporting stalk of a multi-flowered inflorescence or of a solitary flower

rachis-the central axis of an elongate inflorescence

sessile-lacking a pedicel

scape-a leafless flowering stalk

scapose-having a scape

spathe-a comparatively large and sometimes brightly colored bract which partially to completely surrounds an inflorescence

terminal-situated at the apex of a flowering stalk, as opposed to being axillary

The simplest type of inflorescence to imagine, although probably not the simplest from an evolutionary standpoint, is the **solitary** flower which terminates a flowering stalk or which arises from the axil of a leaf. Some authors do not consider the solitary flower as an inflorescence type. In a similar fashion, inflorescences of two to many flowers may be terminal or axillary. Determining the spatial limits of a flower cluster can be difficult. The apex of a terminal inflorescence is obviously the uppermost flower. The lower limit is marked by the appearance of the first well-developed leaf. While much reduced leaves may occur within the inflorescence, especially subtending individual flowers, no leaves of the size and general appearance of ordinary foliage leaves will occur within it. Axillary inflorescences are borne on a floral axis which arises from the axil of a leaf. It is this leaf which marks the lower limit of the axillary inflorescence. In some species, both terminal and axillary flower clusters may occur.

No satisfactory classification of inflorescence types has yet been devised. Past attempts have emphasized either a structural approach or have dwelled upon the blooming sequences within the inflorescence. The scheme offered below is an artificial one, presented solely for ease in determining inflorescence types.

A. Simple inflorescences (not composed of recognizable sub-units which are themselves inflorescence types)

B. Flower solitary

axillary-flower arising from the axil of a leaf along a stem

terminal-situated at the apex of a flowering stalk

B. Flowers two or more per inflorescence

C. No branches within the inflorescence easily discernible

spike- an elongate inflorescence; the flowers sessile, dense or remote from one another

spikelet- a small spike; the flowers inconspicuous and more or less hidden by bracts, as in grasses and sedges

spadix- a spike in which the axis is thick and fleshy; may be associated with an enveloping spathe, as in many members of the philodendron family (Araceae)

catkin (= **ament**)- a pendant or erect inflorescence in which the typically unisexual and apetalous flowers are more or less hidden by scaly bracts; often falls as a single unit; usage traditionally restricted to such woody plants as oaks, willows, birches, and walnuts.

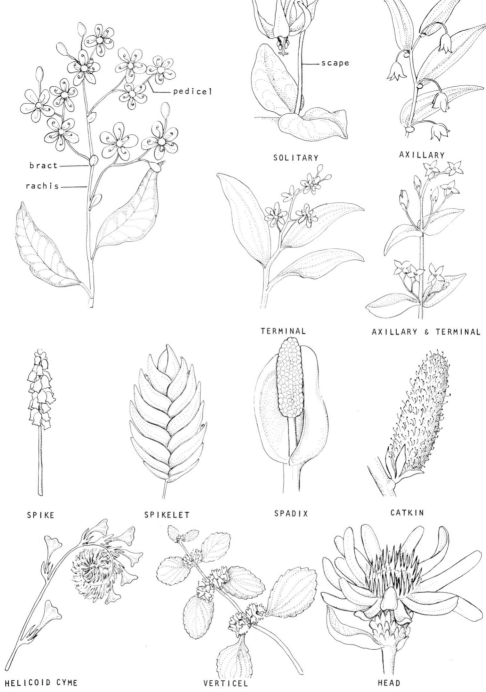

pedicel

scape

bract

rachis

SOLITARY   AXILLARY

TERMINAL   AXILLARY & TERMINAL

SPIKE   SPIKELET   SPADIX   CATKIN

HELICOID CYME   VERTICEL   HEAD

Fig. 17. **Inflorescence types.**

head (= **capitulum**)- a dense spherical or rounded inflorescence of sessile flowers, as in the aster family (Compositae)

helicoid cyme- a one-sided coiled inflorescence resembling a fiddlehead, as in most members of the borage family (Boraginaceae)

verticel- an axillary whorl of flowers radiating in many directions, as in several members of the mint family (Labiatae)

C. Inflorescence with branches easily noted without dissection

raceme- an elongate inflorescence of pedicellate flowers on an unbranched rachis, the flowers dense or remote from one another

umbel- a flat-topped or somewhat rounded inflorescence in which all of the pedicels arise from a common point at the apex of the peduncle, as in the onion and its relatives

corymb- a flat-topped or somewhat rounded inflorescence in which the pedicels of varying length are inserted along the rachis

dichasium (= **simple cyme**)- a cluster of three flowers on a common peduncle, the central flower the oldest and flanked by the others

panicle- a much-branched inflorescence with a central rachis which bears branches which are themselves branched

thyrse- a condensed, often cylindrical or egg-shaped panicle, as in the lilacs

A. Compound inflorescences (composed of simple inflorescence subunits)

compound umbel-a flat-topped or rounded inflorescence in which the peduncles (now called **rays**) of simple umbels (now called **umbellets**) are inserted at the apex of a peduncle, as in most members of the parsley family (Umbelliferae)

raceme of umbels-an elongate inflorescence in which umbels are inserted along a rachis, as in the English ivy

corymb of heads-a corymb in which the branches terminate in heads rather than individual flowers, as in yarrow

panicle of heads-a panicle in which the branchlets terminate in heads rather than individual flowers, as in many members of the aster family (Compositae)

panicle of spikelets-a panicle in which the branchlets terminate in spikelets rather than individual flowers, as in many grasses

spike of spikelets-an inflorescence in which spikelets are sessile along an unbranched rachis

cyme-a flat-topped or rounded inflorescence made up of primary, secondary or tertiary dichasia; the central flower is terminal and is the oldest, as in many members of the stonecrop family (Crassulaceae)

Inflorescence types may also be classified by the sequence of flowering within a flower cluster, with special attention being paid to the location of the oldest flower. In **racemose** or **indeterminate** inflorescences, the blooming sequence is from the base to the top of an

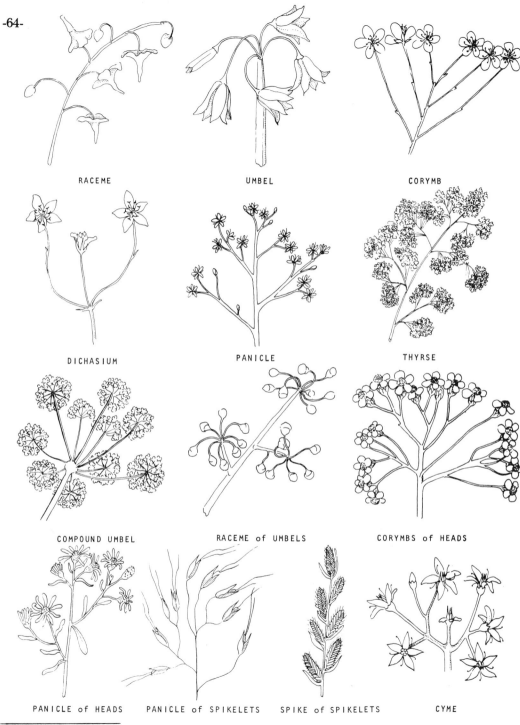

RACEME   UMBEL   CORYMB

DICHASIUM   PANICLE   THYRSE

COMPOUND UMBEL   RACEME of UMBELS   CORYMBS of HEADS

PANICLE of HEADS   PANICLE of SPIKELETS   SPIKE of SPIKELETS   CYME

Fig. 18. **Inflorescence types.**

elongate flower cluster or from the outside toward the center if it is flat-topped. The oldest flower is at the base or outside of the inflorescence, respectively. Spikes, racemes, panicles, corymbs, umbels, and heads often show a racemose blooming sequence.

In **cymose** or **determinate** inflorescences, the blooming sequence is from the top downward or from the center outward. The oldest flower is at the apex or the center of the flower cluster. The dichasium and various cymes often show this sequence.

These racemose or cymose sequences were once thought to be critical in determining fundamental differences in plants. More recently investigators have suggested that they are not nearly so basic and that relationships can be obscured by convergent evolution. The umbel, for instance, may be either racemose or cymose. Several of the more complex inflorescences are mixtures of the two. The evolutionary relationships among the various inflorescence types are indeed obscure.

FRUIT TYPES

MORPHOLOGY

A fruit may be defined as a ripened ovary and its contents, along with any other floral or vegetative parts which might be attached and which mature along with it. Many taxonomists restrict the use of the term "fruit" to the flowering plants and do not refer to the matured female reproductive structures in gymnosperms as fruits. The botanical definition does not carry over into common usage. Sunflower, corn, and carrot "seeds" are actually the fruits of these plants. Fruits such as the squash, bean, pea, and eggplant are called "vegetables".

The following terms are applied to fruits and their parts:

carpophore-an extension of the floral axis between adjacent carpels, as in the fruits of the parsley family

cupule-a series of fused bracts which form a cup beneath the true fruit, as in the acorn

dehiscent-a fruit which opens by sutures, pores or caps

didymous-a strongly lobed fruit, thus appearing as a pair

endocarp-the innermost layer of the fruit wall; it may be soft, papery or bony

exocarp-the outermost layer of the fruit wall; it may be the "skin" of the fruit, a leathery rind or quite hard

false fruit-the structure resulting from the coalescence of the separate fruits of an apocarpous gynoecium, as in the raspberry or from the coalescence of the fruits from an entire inflorescence, as in the pineapple

funiculus-the seed stalk

indehiscent-a fruit which does not open by sutures, pores or caps, the seeds being released by the rotting away of the fruit wall

involucre-a set of separate or fused bracts associated with a fruit, as in the walnut

longitudinal-lengthwise

mesocarp-the middle layer of the fruit wall; often the fleshy edible portion

pericarp-the fruit wall, made up of endocarp, mesocarp, and exocarp

replum-an internal partition (septum) which persists after dehiscence, as in the mustard fruits

septum-an internal partition within the fruit

suture-a line along which a dehiscent fruit opens

transverse-crosswise

true fruit-those fruits derived from the syncarpous or apocarpous gynoecium of a single flower

valve-a segment of a dehisced fruit

A CLASSIFICATION OF FRUIT TYPES

There are two ways of approaching the classification and naming of fruits. One is essentially morphological. Fruits are distinguished by their fleshiness, dehiscence, carpel number, etc. The second view is basically taxonomic. A particular fruit is given a name because of the family to which the plant belongs, with little regard to certain details of its structure. Following this view, the legume is the fruit of the Leguminosae, regardless of its general appearance. The scheme presented below is an artificial grouping of the traditionally recognized fruit types.

A. Dry fruits
  B. Fruits 1-seeded and indehiscent

   achene (=**akene**)-seed and pericarp attached only at the funiculus, the seed usually tightly enclosed by the fruit wall, as in the sunflower and buckwheat

   cypsela-an achene with an adnate calyx, as in the aster family

   utricle-a small, bladdery achene-like fruit with the seed loosely surrounded by the fruit wall, as in the pigweed

   caryopsis (=**grain**)-seed and pericarp completely fused, as in the grass family

   samara-a winged achene, as in the elms and ashes

   nut-derived from a syncarpous gynoecium, but 1-seeded by abortion of carpels; exocarp usually hard; the fruit often subtended by an involucre; walnut, acorn

   nutlet-a small nut

   schizocarp-derived from a syncarpous gynoecium, the carpels separating from one another into 1-seeded indehiscent segments, as in the mallow and parsley family

B. Fruits 2-to many-seeded and dehiscent

   capsule-derived exclusively from a syncarpous gynoecium; typically several-to many-seeded (rarely 1-seeded); several types are recognized:

    poricidal-opening by a series of pores near the top, as in the poppy

    pyxis (=**circumscissile**)-opening by a lid, as in the purslane

    denticidal-opening apically, leaving a ring of teeth, as in the chickweeds

    loculicidal-dehiscing lengthwise, the sutures opening within a locule, as in the iris

    septicidal-dehiscing lengthwise, the sutures splitting a septum, as in the yucca

   silique-gynoecium bicarpellate, the walls peeling away from a papery central partition (replum); the fruit type of the mustard family

   silicle-a silique which is not more than 2-3 longer than wide

   legume-unicarpellate, dehiscing along both sutures; the fruit type of the pea family

   loment-a legume with pronounced constrictions between the seeds; dehiscing transversely between the seeds

   follicle-unicarpellate, dehiscing along one suture; note that a follicle may be found in the apocarpous gynoecium (each carpel forming one, as in the rose and magnolia family) or the syncarpous gynoecium by means of the separation of carpels at maturity, as in the milkweeds

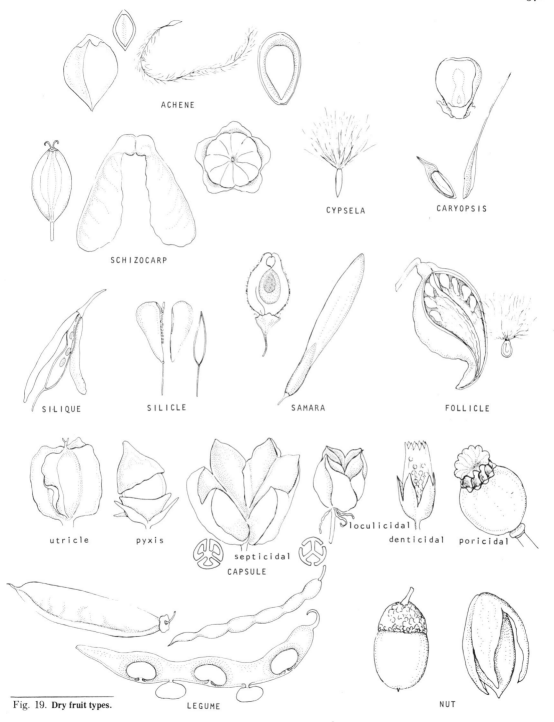

ACHENE

CYPSELA

CARYOPSIS

SCHIZOCARP

SILIQUE   SILICLE   SAMARA   FOLLICLE

utricle   pyxis   septicidal   loculicidal   denticidal   poricidal
                            CAPSULE

LEGUME   NUT

Fig. 19. **Dry fruit types.**

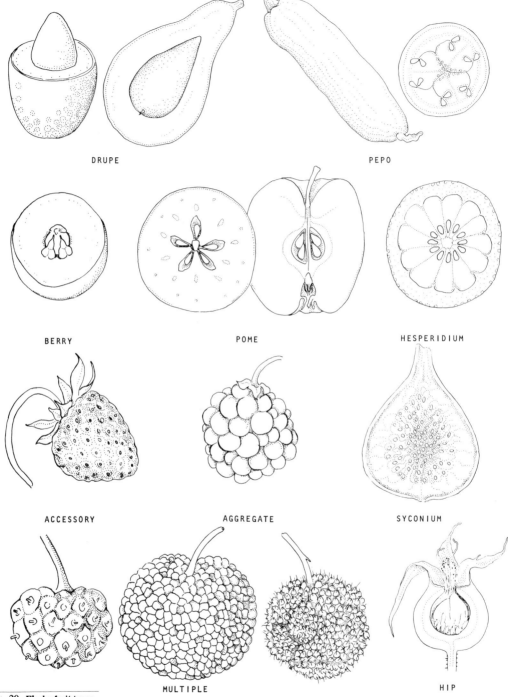

DRUPE

PEPO

BERRY

POME

HESPERIDIUM

ACCESSORY

AGGREGATE

SYCONIUM

MULTIPLE

HIP

Fig. 20. **Fleshy fruit types.**

A. Fleshy fruits

    C. "True fruits" (derived from a syncarpous gynoecium of a single flower)

        drupe-exocarp the "skin"; mesocarp fleshy; endocarp bony; the seed and endocarp constitute a **pyrene;** peach

        berry-entire pericarp soft, as in the tomato or grape

        pepo-a berry with a leathery rind; derived from an inferior ovary; use often restricted to the squash family

        pome-ovary inferior, surrounded by fleshy tissue usually interpreted as a hypanthium, as in the apple or pear

        hesperidium-ovary superior; septations conspicuous, these lined with fleshy hairs; restricted to the citrus fruits

    C. "False fruits" (the structures resulting from the coalescing of the separate true fruits of an apocarpous gynoecium or the coalescing of the fruits from an entire inflorescence)

        accessory-fruit formed from the expanded dome-like receptacle of a single flower; covered with numerous achenes; known only in the strawberry and false strawberry

        aggregate-fruit formed from the many separate dry or fleshy fruits of a single flower, as in the raspberry or magnolia

        hip-a vase-like leathery hypanthium containing several achenes; restricted to the rose

        multiple-fruit derived from the fusion of an entire inflorescence, as in the pineapple or sycamore

        syconium-a hollow, vase-like inflorescence with the flowers lining the inside; restricted to the fig

A KEY TO FRUIT TYPES

1.   Fruit derived from a single flower

    2. Fruit formed from a hollow leathery hypanthium which contains several to many seed-like achenes . . . . **hip**

    2. Fruit dry or fleshy, but not hollow, leathery, and containing achenes

3. Fruit formed from the few to many separate carpels of an apocarpous gynoecium, these often appearing partially fused at maturity

    4. Fruit an enlarged fleshy receptacle covered with tiny seed-like achenes . . .**accessory**

    4. Bulk of the fruit consisting of true ovary tissue

        5. Gynoecium of 2-several separate carpels, each opening along a single suture . . . . **follicle**
        5. Gynoecium of many separate carpels (fleshy or dry), often partially fused at maturity . . . . **aggregate**

3. Fruit derived from a syncarpous or unicarpellate gynoecium

    6. Fruit dry at maturity

        7. Fruit indehiscent and 1-seeded

            8. Fruit winged . . . . **samara**

            8. Fruit wingless

                9. Fruit with a leathery to hard bony outer wall, often partially to complete-ly enveloped by an involucre . . . . **nut**

            9. Fruit typically with a thin papery outer wall; involucre usually lacking

                10. Fruit splitting into two or more single-seeded components, each remaining closed . . . . **schizocarp**

                10. Fruits 1-seeded, indehiscent

                    11. Calyx persistent, adnate to the apex of the fruit . . . .**cypsela**

                    11. Calyx deciduous

                        12. Seed completely fused to the fruit wall . . . .**caryopsis**

                        12. Seed and fruit wall separable

                            13. Fruit bladdery, the seed loose . . . . **utricle**

                            13. Fruit not bladdery, its wall tight about the seed . . . **achene**

      7. Fruits dehiscent, typically with two or more seeds

14. Fruit dehiscing apically by a series of small pores, teeth or a detachable lid

    15. Fruit opening by means of a detachable lid . . . . **pyxis (=circumscissile capsule)**

    15. Fruit opening by means of pores or teeth

        16. Fruit dehiscing by a series of teeth . . . . **denticidal capsule**

        16. Fruit dehiscing by a series of small pores . . . . **poricidal capsule**

14. Fruit dehiscing longitudinally into halves or transversely into segments

    17. Fruit composed of a single carpel

        18. Fruit with prominent constrictions between the seeds, dehiscing transversely into 1-seeded segments . . . . **loment**

        18. Fruits without prominent constrictions, dehiscing longitudinally

            19. Fruit opening along two sutures . . . . **legume**

            19. Fruit opening along one suture . . . . **follicle**

    17. Fruit composed of two or more carpels

        20. Carpels (not fruits) 1-seeded . . . . **schizocarp**

        20. Carpels (not fruits) several-to many-seeded

            21. Fruit bicarpellate, the two carpel walls separating and falling away from a papery central partition at maturity

                22. Fruit three times longer than wide . . . . **silique**

                22. Fruit less than twice as long as wide . . . . **silicle**

            21. Fruit composed of two or more carpels, but not dehiscing as above

                23. Sutures opening within a locule . . . . **loculicidal capsule**

                23. Sutures splitting through the septum . . . . **septicidal capsule**

   6. Fruit fleshy at maturity

      24. Fruit 1-seeded with a bony endocarp ....**drupe**

      24. Fruit multi-seeded; endocarp fleshy or papery, but not bony

         25. Fruit derived from a superior ovary

            26. Fruit with evident segments, divided by fibrous or papery walls lined with juicy hairs.......................................**hesperidium**

            26. Fruit not conspicuously segmented; interior walls, if present, not lined with juicy hairs ....**berry**

         25. Fruit derived from an inferior ovary

            27. Fruit with a leathery exocarp .... **pepo**

            27. Fruit with a thin skin

               28. Fruit wall fleshy throughout .... **berry**

               28. Fruit wall with papery endocarps .... **pome**

1. Fruit derived from an entire inflorescence (often with a warty or faceted exterior)

   29. Fruit vase-like and hollow when viewed in longitudinal section; smooth exterior.**syconium**

   29. Fruit solid; exterior warty or faceted .... **multiple**

6- A Survey of the Flowering Plants

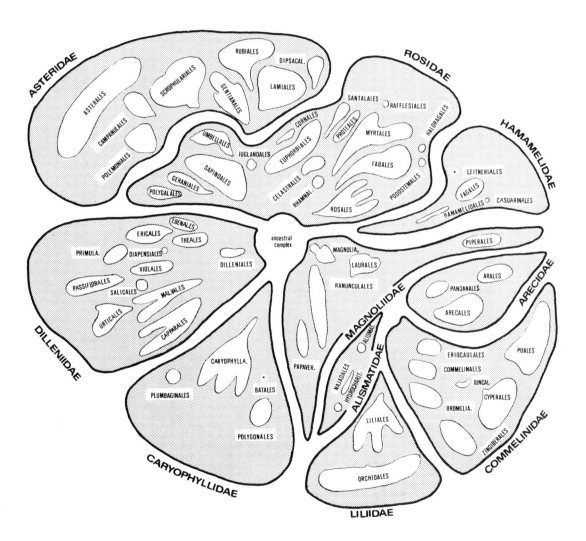

Fig. 21. The phylogenetic relationships of selected orders of flowering plants (After Stebbins, 1974).

Magnoliopsida

Order: Family	Common Name	Genera/Species	Distribution	Form	K
Magnoliales					
Magnoliaceae	Magnolia	12/230	OW & NW; temp.	TS	3
Annonaceae	Custard apple	120/2100	Trop.	TS	3[3+3]
Canellaceae	Wild cinnamon	5/16	OW & NW Trop.	T	4-5
Illiciaceae	Illicium	1/42	OW & NW	ST	x-∞
Calycanthaceae	Calycanthus	2/6	E. As. & N. Am.	S	x
Lauraceae	Laurel	32/2000-2500	Trop. & subtrop.	TS	N
Piperales					
Saururaceae	Lizard's tail	5/7	E. As. & N. Am.	H	0
Piperaceae	Pepper	10-12/3000	Trop.	HSTV	0
Aristolochiales					
Aristolochiaceae	Birthwort	7/400	Trop. & temp.	HV	(3)
Nymphaeales					
Nymphaeaceae	Water lily	7/68	Cosmo.	H	3-∞
Nelumbonaceae	Indian lotus	1/2	Temp. & trop.	H	4-5
Ceratophyllaceae	Hornwort	1/10	Cosmo.	H	(9-12)
Ranunculales					
Ranunculaceae	Buttercup	35-70/2000	OW & NW; temp.	HS[V]	3-x
Berberidaceae	Barberry	9/590	N. hemis. & S. Am.	S[H]	3+3
Podophyllaceae	Mayapple	8/25	Temp. n. hemis.	H	4-x
Menispermaceae	Moonseed	65/350	OW & NW; temp.	V[ST]	3+3
Papaverales					
Papaveraceae	Poppy	26/200	Temp. & subtrop.	H[S]	2-3
Fumariaceae	Fumitory	16/450	N. temp.	H	2

agnoliidae

C	A	G	Fruit Type(s)	Miscellaneous Comments
6-∞	∞	∞	fol, sam, agg	Sheathing stipules; showy flowers
3	∞	∞	ber, agg	*Asimina* occurs in eastern U.S.
4-12	20-40	(2-5)	ber	*Canella* in S. Florida
x-∞	4-∞	5-20	fol	Southeastern U.S.
x	5-30	∞	ach	California & Southeastern U.S.
N	4N	N	ber	N usually = 3
0	3,6,8	3-4	fol, cap	*Saururus* and *Anemopsis* native
0	1-10	1-4	ber	Source of black and white pepper
0	6-36	(4-6)	cap	*Asarum* and *Aristolochia* native
3-∞	∞	∞	fol, nutl, ber	Showy perennial aquatics
∞	∞	∞	nutl	Fruits in flat-topped receptacle
0	12-16	1	ach	Submerged unisexual aquatics
0-∞	∞	∞	fol, ach, ber	Often with dissected leaves
3+3	4-18	(2-3)	ber	Stamens dehiscing by flap-covered valves
6-9	4-x	1	ber, fol	Traditionally included in Berberidaceae
3+3	3+3	3	drp, ach	Three genera native to U.S.
4-12	∞	(2-x)	cap	Sepals caducous; often with brightly colored latex
2+2z	4-6	(2)	cap, nut	Sap watery

The Magnoliidae consist of six orders, 36 families, and more than 11,000 species. The subclass is the most primitive of the Magnoliopsida. No combination of simple features defines the group. Typically, the flowers have well-developed perianths, although they are not necessarily differentiated into a calyx and corolla. The androecium is typically of numerous stamens. The gynoecium is apocarpous. Some plants have ethereal oil cells, specialized secretory cells.

The orders may be summarized as follows. The Magnoliales are the most primitive order of the subclass. It is here that we find the flowering plants which retain such features as open carpels and vesselless wood. The group has been called the "Magnolian Alliance" by some workers. The Piperales are herbaceous plants with reduced flowers and dense inflorescences. The Aristolochiales are a combination of primitive and advanced features. They may be epigynous and are usually apetalous, with a petaloid calyx. The Nymphaeales are vesselless aquatics. The delimitation of families remains controversial. The Ranunculales (= Ranales of older authors) are herbaceous counterparts of the Magnoliales, differing from them in lacking the very primitive features associated with the Magnolian Alliance (open carpels, vesselless wood, etc.). Many of the species are found in wet or moist habitats. The Papaverales consist of the Papaveraceae and Fumariaceae, often treated as a single family in older literature. The plants typically have compound leaves and their flowers are perfect, hypogynous, and syncarpous. Latex and isoquinoline alkaloids are often present.

MAGNOLIACEAE
Magnolia Family

Trees and shrubs. Leaves alternate, simple, deciduous or evergreen, stipulate, the stipules large and enclosing the young buds, falling quickly and leaving an annular scar at the node. Flowers perfect, often large and showy, actinomorphic, solitary, with deciduous spathaceous bracts; receptacle elongate. Perianth of separate sepals and petals, although these are not always differentiated. Calyx typically of three sepals. Corolla of six to many petals. Androecium of numerous separate stamens, spirally inserted. Gynoecium apocarpous, of many (rarely one to three) spirally inserted carpels. Fruit a follicle, samara, rarely a berry.

12/230; temperate regions of the northern hemisphere, with centers of distribution in eastern Asia, Malaysia, and eastern North America. Magnolia and Liriodendron are native to the U.S. Economic products include wood for furniture and cabinets; several species are ornamentals.

Selected Genera:

Magnolia (77)-magnolia, bull bay, cucumber tree

Michelia (15)-a widely planted ornamental

Liriodendron (2)-L. tulipifera is the tulip tree or yellow poplar

Recognition Characters:

Trees or shrubs with conspicuous, actinomorphic flowers with numerous stamens and carpels.

$$K^3 C^{6-\infty} A^\infty \underline{G}^\infty$$

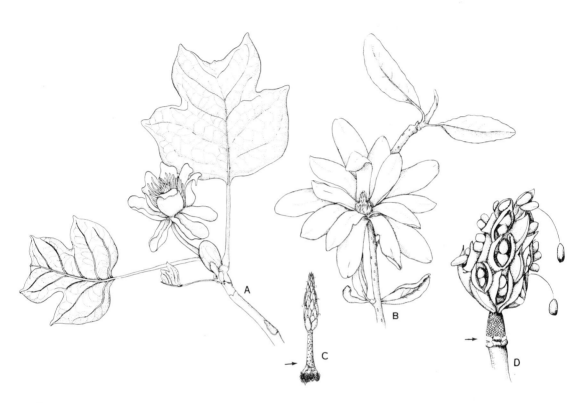

Fig. 22. **Magnoliaceae**. a, *Liriodendron;* b-d, *Magnolia* flower, floral axis without perianth and androecium (arrows indicate their points of attachment), and gynoecium at fruiting stage (follicles have dehisced to reveal seeds).

NYMPHAEACEAE
Water Lily Family

Perennial aquatics with erect stems or creeping rhizomes. Leaves simple, alternate, floating or emersed, orbicular or cordate. Flowers bisexual, solitary, conspicuous, actinomorphic, often fragrant. Calyx of three to many separate sepals, these often green. Corolla of three to many separate petals, usually with nectaries. Androecium of three to many separate stamens. Gynoecium apocarpous, of three to many carpels (sometimes syncarpous, rarely unicarpellate), the ovary superior. Fruit a follicle or berry.

7/68; throughout the world in aquatic habitats. Four genera are native to the U.S. A source of food for birds and aquatic animals; many species of *Nymphaea* are in cultivation. The family, as treated here, includes the Euryalaceae, Barclayaceae, and the Cabombaceae. All of these have been segregated from the traditional Nymphaeaceae by recent workers.

Selected Genera:

Subfamily: Cabomboideae (=Cabombaceae)

Cabomba (7)-fanwort, water shield

Brasenia (1)-*B. schreberi* is the water shield

Subfamily: Nymphaeoideae (=Nymphaeaceae in the strict sense)

Nuphar (7-20)-yellow water lily

Nymphaea (40)-water lily

Victoria (2)-*V. amazonica* is the giant Amazon water lily

Recognition Characters:

Perennial freshwater aquatics with large leaves and conspicuous flowers. The flowers actinomorphic with numerous perianth parts, stamens, and carpels.

$$K^{3-\infty} C^{3-\infty} A^{\infty} \underline{G}^{\infty}$$

NELUMBONACEAE
Lotus Family

Perennial aquatic herbs from creeping rhizomes. Leaves simple, alternate, peltate, suborbicular, and usually elevated above the water on stout petioles. Flowers solitary, conspicuous, and bisexual. Calyx of 4 or 5 separate sepals. Corolla of many separate petals. Androecium of many separate stamens. Gynoecium of many carpels, separately inserted in a flat-topped receptacle, the ovary superior. Fruit a nutlet.

1/2; eastern U.S. to S. America, Asia, and Australia. *Nelumbo pentapetala* is native to the U.S. The other species, *N. nucifera* is the sacred lotus of the Old World. Both species are cultivated as ornamentals.

Genus:

Nelumbo (2)-lotus, American lotus, sacred lotus

Recognition Characters:

Large aquatic herbs with suborbicular peltate leaves elevated above the water. Flowers large; perianth parts, stamens, and carpels numerous, the receptacle large and flat-topped.

$$K^{4-5} C^{\infty} A^{\infty} \underline{G}^{\infty}$$

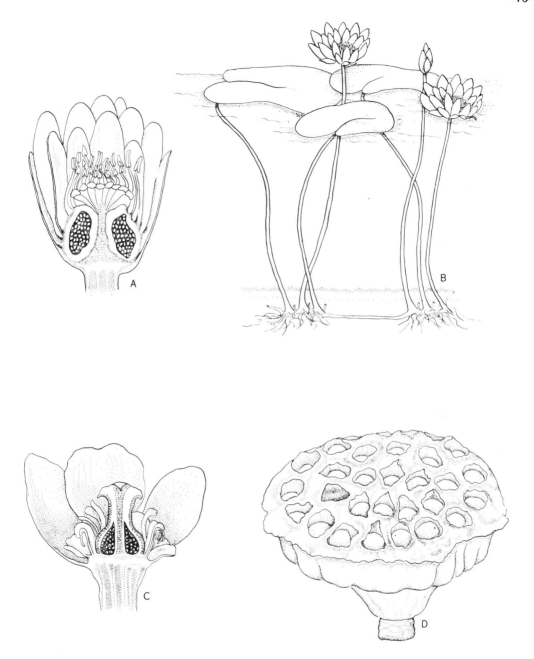

Fig. 23. **Nymphaeaceae.** a & b, *Nymphaea* flower in longitudinal section and habit; c, *Nuphar* flower in longtiudinal section. **Nelumbonaceae.** d, *Nelumbo* gynoecium showing seeds in cavities in flat-topped receptacle.

RANUNCULACEAE
Buttercup or Crowfoot Family

Annual or perennial herbs, sometimes woody. Leaves usually alternate (opposite in *Clematis* and certain *Ranunculus*), palmately compound, exstipulate, with a sheathing leaf base. Flowers solitary or in racemes or cymes; bisexual (rarely unisexual); actinomorphic or zygomorphic. Floral series inserted on an elongate receptacle, distinct, variable in number. Perianth usually petaloid, often not differentiated into a true calyx and corolla; commonly yellow, white, or blue. Stamens usually numerous, spirally inserted. Gynoecium apocarpous, of three to many carpels (rarely syncarpous or unicarpellate), the ovary superior. Fruit a follicle, achene, berry, or rarely a capsule.

35-70/2000; cooler regions of the northern hemisphere. Twenty-one genera are native to the U.S. The family is the source of many ornamentals, a few drug plants, and important toxic plants, such as *Delphinium*.

Selected Genera:

Subfamily: Helleboroideae (fruit a follicle, capsule, or berry)

Helleborus (20-25)-hellebore

Caltha (20)-marsh marigold

Aquilegia (70)-columbine; many in cultivation

Delphinium (300)-larkspur; many are important livestock poisoners in the west

Aconitum (275-300)-monk's hood; several are toxic

Subfamily: Ranunculoideae (fruit an achene)

Ranunculus (400)-buttercup; widespread

Thalictrum (120)-meadow rue; unisexual flowers

Anemone (120)-anemone, wind flower, pasque flower

Clematis (300)-virgin's bower; leather flower; vine with opposite leaves

Recognition Characters:

Herbs with compound leaves with sheathing leaf bases. Flowers with numerous stamens and carpels, typically actinomorphic, but zygomorphic in *Delphinium* and *Aconitum*.

$$K^{3-x} \; C^{0-\infty} \; A^{\infty} \; \underline{G}^{\infty \; [3 \text{ or } 1]}$$

Fig. 24. **Ranunculaceae.** a-c, *Ranunculus* fruits, plant, and flower; d, *Aquilegia;* e, *Delphinium,* exploded dissection of flower (arrows indicate sepals). (Drawing d after Porter, 1967.)

BERBERIDACEAE
Barberry Family

Shrubs or perennial herbs. Leaves alternate, simple or pinnately compound, persistent or deciduous, mostly exstipulate, often spiny. Flowers in cymes, racemes, fascicles or solitary; actinomorphic; bisexual. Calyx of 4-6 separate sepals. Corolla of 4-6 separate petals, the perianth often appearing 3+3+3+3. Androecium of 4-18 biseriate, pressure sensitive stamens, these dehiscing by flap-like valves. Gynoecium syncarpous, of 2-3 carpels, appearing unicarpellate and unilocular. Fruit a berry.

9/590; distributed throughout the northern hemisphere and South America. *Berberis* and *Mahonia* are native to the U.S. *Berberis vulgaris* is the alternate host of the wheat rust, *Puccinia graminis*. Several species are cultivated as ornamentals. As treated here, the family includes the Nandinaceae and Leonticaceae.

Selected Genera:

Berberis (450)-barberry; leaves simple, branches spiny

Mahonia (70)-barberry; leaves pinnately compound, branches unarmed

Nandina (1)-*N. domestica* is the heavenly bamboo

Recognition Characters:

Ours shrubs with biseriate stamens, dehiscing by flaps.

$$K^{3+3} \ C^{3+3} A^{4-18} \underline{G}^{(2-3)}$$

PODOPHYLLACEAE
Mayapple Family

Perennial herbs with fleshy rootstocks. Leaves palmately lobed or 2-3 foliolate, spineless. Flowers actinomorphic, bisexual. Calyx of 4-15 separate sepals. Corolla of 6-9 separate petals, rarely with no perianth *(Achlys)*. Androecium of 4-18 stamens, as many as or twice as many as the number of petals, the anthers opening by slits or valves. Gynoecium unicarpellate and unilocular, the ovary superior. Fruit a berry or follicle.

8/25; temperate northern hemisphere. The 5 genera listed below are native to the U.S. Of little economic importance; *Podophyllum* yields a mitosis inhibitor. Its leaves and roots are toxic.

Selected Genera:

Podophyllum (6)-mayapple, American mandrake

Jeffersonia (1)-*J. diphylla* is the twin leaf

Achlys (2)-vanilla leaf

Vancouveria (3)-inside-out-flower

Caulophyllum (2)-cohosh, blue cohosh

Recognition Features:

Herbs with palmately lobed or 2-3 foliolate leaves; gynoecium unicarpellate and unilocular.

$$K^{4-x} C^{6-9} A^{4-x} \underline{G}^1$$

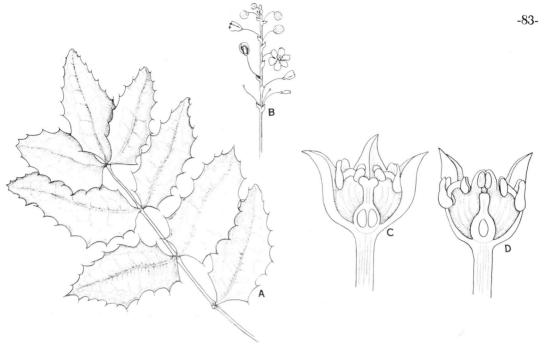

Fig. 25. **Berberidaceae.** a-d, *Mahonia* leaf, inflorescence, and two views of a flower in longitudinal section.

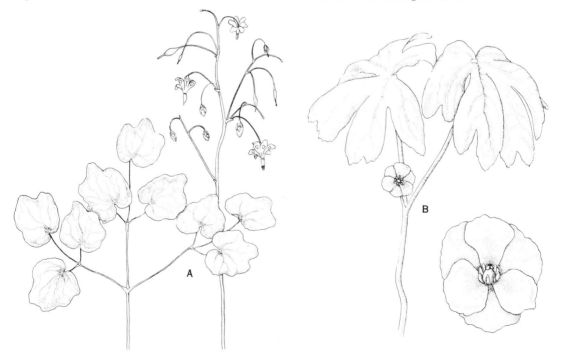

Fig. 26. **Podophyllaceae.** a, *Vancouveria;* b & c, *Podophyllum* plant and flower detail.

PAPAVERACEAE
Poppy Family

Mostly annual or perennial herbs, sometimes shrubs, rarely trees, **often** with milky or colored latex. Leaves alternate, entire or variously divided. Flowers showy, bisexual, actinomorphic. Calyx of 2-3 distinct, quickly falling sepals (these united in *Eschscholzia*). Corolla of 4-6 or 8-12 biseriate, crumpled, separate petals. Androecium of numerous stamens. Gynoecium of two to several united carpels, unilocular or several loculed with intrusive placentae, ovary superior. Fruit a capsule, opening by valves or pores.

26/200; mostly subtropical and temperate areas of the northern hemisphere, particularly western North America. Thirteen genera are native to the U.S.

Of little economic importance except for *Papaver somniferum* which yields opium and its many derivatives, including morphine and heroin; a few genera are cultivated as ornamentals.

Selected Genera:

Papaver (100)-poppy; the source of opium and many ornamentals

Sanguinaria (1)-blood root

Eschscholzia (123)-*E. californica* is the California poppy

Argemone (9)-prickly poppy

Platystemon (57)-cream cups

Dendromecon (20)-tree poppy; shrubby

Recognition Characters:

Herbs with deciduous sepals, crumpled petals, numerous stamens, and a milky or colored latex (when present!).

$$K^{2-3} C^{4-12} A^{\infty} \underline{G}^{(2-x)}$$

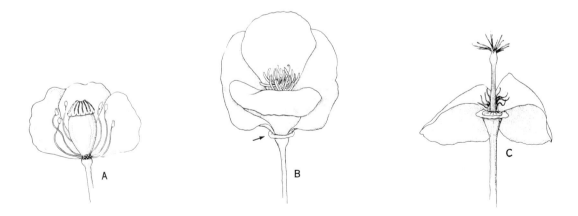

Fig. 27. **Papaveraceae.** a, *Papaver* flower with some petals missing to reveal gynoecium; b & c, *Eschscholzia* flower (arrow indicates expanded rim around the receptacle).

FUMARIACEAE
Dutchman's Breeches Family

Herbs with brittle stems and watery juice. Leaves alternate, either cauline or in rosettes, usually much-dissected. Flowers zygomorphic, bisexual, usually in racemes. Calyx of two minute, deciduous sepals. Corolla of four petals, in a 2+2 configuration, the outer pair with basal sacs or spurs; the inner pair narrower and coherent at the apex. Androecium of six stamens, united into two sets of three on either side of the gynoecium; or four stamens, free and opposite the petals. Gynoecium bicarpellate, syncarpous, unilocular, ovary superior. Fruit a septate capsule or a 1-seeded nut.

16/450; chiefly north temperate regions of the Old World. *Adlumia*, *Dicentra*, and *Corydalis* are native to the U.S. Of little economic importance; a few are cultivated as ornamentals.

Selected Genera:

Fumaria (50)- fumitory

Dicentra (17)- dutchman's breeches; bleeding hearts

Corydalis (280)- corydalis

Recognition Characters:

Herbs with dissected leaves. Flowers zygomorphic; sepals deciduous; petals paired, outer sac-like. Sometimes considered as part of the Papaveraceae, with which it is compared in Table 6 below.

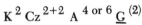

$$K^2 \, Cz^{2+2} \, A^{4 \text{ or } 6} \, \underline{G}^{(2)}$$

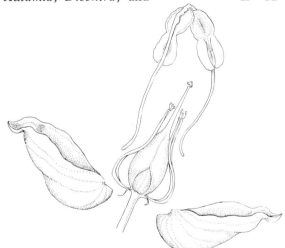

Fig. 28. **Fumariaceae.** Exploded dissection of *Dicentra* showing inner and outer pairs of petals.

Table 6. Comparison of the Papaveraceae and Fumariaceae.

Character	Papaveraceae	Fumariaceae
sap	milky	watery
symmetry	actinomorphic	zygomorphic
stamen number	numerous	4 or 6

Magnoliopsida

Order: Family	Common Name	Genera/Species	Distribution	Form	K
Hamamelidales					
Platanaceae	Plane Tree	1/10	N. temp & subtrop.	T	4?
Hamamelidaceae	Witch Hazel	25/100	Subtrop.	TS	4-5
Urticales					
Ulmaceae	Elm	15/200	Temp. & subtrop.	TS	4-8
Moraceae	Mulberry	53/1400	Pantropical	TS	4
Cannabaceae	Hemp	2/3-5	N. temp.	HV	5
Urticaceae	Nettle	45/550	Trop. & temp.	HS	4-5
Leitneriales					
Leitneriaceae	Cork Wood	1/1	SE U.S.	S	0?
Juglandales					
Juglandaceae	Walnut	7/50	N. temp.	T	3-6
Myricales					
Myricaceae	Sweet Gale	1-4/40	Cosmo.	TS	0
Fagales					
Fagaceae	Beech	8/900	Cosmo.	TS	4-6
Betulaceae	Birch	6/150	N. hemis.	TS	0or4
Casuarinales					
Casuarinaceae	Beef Wood	1-2/65	S. hemis.	TS	0

amamelidae

C	A	G	Fruit Type(s)	Miscellaneous Comments
4?	3-4	<u>6-9</u>	agg of ach	*Platanus* is the sycamore
4-5	4-5	<u>(2)</u>	cap	Often with stellate hairs
0	4-8	(2)	sam, nut, drp	Oblique leaf bases
0	4	(2)	ach,nut,drp,syc	Milky latex often present
0	5	<u>1</u>	ach	Family of hops and marijuana
0	4-5	<u>1</u>	ach, drp	Only one tribe has stinging hairs!
0	3-12	<u>1</u>	drp	*Leitneria* is endemic to the U.S.
0	3-∞	$\overline{(2\text{-}3)}$	drp or nut	Leaves pinnately compound
0	2-x	(2)	drp	Aromatic, dotted leaves
0	4-∞	$\overline{(3)}$	nut	Fruit ± enclosed by a cupule
0	2-20	$\overline{(2)}$	nut, sam	Much variation in pistillate catkin
0	1	<u>(2)</u>	nut	*Equisetum*-like foliage

With its nine orders, 23 families, and 3400 species, the Hamamelidae are the smallest of the subclasses within the Magnoliopsida. More than half the families have fewer than ten species; five contain only one. The group is composed primarily of woody plants with reduced anemophilous flowers. The more advanced members have unisexual flowers born in catkins and have indehiscent, 1 - seeded, unilocular fruits. These plants constitute the core of a group referred to as the "Amentiferae" by Adolf Engler. The name literally means "catkin-bearing", a reference to the typical inflorescence of these plants. The Amentiferae were considered by Engler and his followers as constituting a natural group of closely related species which stood at the base of the dicotyledonous families. Most, if not all, present day students of plant families have rejected both the naturalness and the primitiveness of the Amentiferae.

The Urticales are an exception to the woodiness of the subclass. Within the order the Moraceae and Urticaceae are quite close, in fact, separated only with some difficulty. The Leitneriales are represented in the U. S. by a single species in the Southeast. It is isolated and of uncertain relationships. The Casuarinales consist of about fifty species of the genus *Casuarina*. Although none is native, *C. equisetifolia* is a common ornamental tree in the warmer parts of the country.

PLATANACEAE
Plane Tree or Sycamore Family

Large trees with deciduous bark. Leaves simple, alternate, palmately lobed or nerved with dilated petiole base concealing the bud. Flowers unisexual, the species monoecious. Flowers borne in pendulous, spherical heads. Details of the flower structure remain controversial, some authors claiming a 4-merous basic pattern, while others maintain that the flower is without any perianth. Fruit an aggregate of achenes.

1/10; temperate and subtropical northern hemisphere. A few species are cultivated as ornamentals.

Genus:

Platanus (10)- sycamore, plane tree

Recognition Characters:

Trees with peeling bark, palmately lobed and veined leaves, the dilated petiole base hiding the bud; spherical heads of unisexual flowers.

Fig. 29. **Platanaceae.** a & b, *Platanus* leaves and inflorescence.

HAMAMELIDACEAE
Witch Hazel Family

Trees and shrubs, often with stellate hairs. Leaves alternate, rarely opposite, deciduous or evergreen, simple, sometimes glandular toothed, stipulate. Flowers bisexual or unisexual (the species monoecious), small, actinomorphic or zygomorphic. Calyx of 4-5[0] basally fused sepals. Corolla of 4-5[0] separate petals. Androecium of 4-5 stamens. Gynoecium bicarpellate, the two carpels often free at the tips, ovary inferior or rarely superior, the flowers thus epigynous or hypogynous. Fruit a loculicidal capsule, often with a hard exocarp and endocarp.

25/100; found chiefly in the subtropical areas of Africa, Asia, and Australia. *Hamamelis, Liquidambar,* and *Fothergilla* are native to the U.S. Of economic importance as a source of timber, resins, liniments (hamamelis bark), and ornamentals.

Selected Genera:

Hamamelis (8)- witch hazel

Fothergilla (4)- fothergilla; Southeastern states

Liquidambar (5)- *L. styraciflua* is the sweet gum

Recognition Characters:

Woody plants with stellate hairs, bicarpellate gynoecium with divergent apices, and a woody or leathery capsule.

$$K \underset{\smile}{4\text{-}5} \ C \ 4\text{-}5 \ A \ 4\text{-}5 \ \overline{G} \ (2)$$

ULMACEAE
Elm Family

Trees and shrubs. Leaves simple, alternate, often with oblique bases, stipules paired. Flowers bisexual (*Ulmus*) or unisexual; zygomorphic. Calyx of 4-8 sepals. Corolla absent. Androecium of 4-8 stamens. Gynoecium bicarpellate, superior, styles 2. Fruit a samara, nut, or drupe.

15/200; temperate and subtropical regions of the northern hemisphere. *Ulmus, Planera,* and *Celtis* are native to the U.S. Of economic importance as a source of wood. *Ulmus americana,* widely planted as an ornamental, is now suffering from the Dutch elm disease.

Selected Genera:

Subfamily: Ulmoideae (hypanthium present)

Ulmus (ca. 20)- elm; east of the Rockies

Planera (1)- *P. abelica* is the planer tree; Southeastern

Subfamily: Celtoideae (hypanthium absent)

Celtis (70)- hackberry

Zelkova (4)- an introduced ornamental

Recognition Characters:

Trees or shrubs with oblique leaf bases, often with offset buds, fruit an evenly winged samara or drupe.

$$K \ 4\text{-}8 \ C \ 0 \ A \ 4\text{-}8 \ \underline{G} \ (2)$$

Fig. 30. **Hamamelidaceae.** a, *Liquidambar*; b & c, *Hamamelis* fruiting branch and flower. **Ulmaceae.** d & e, *Ulmus* leaves and fruits; f, *Celtis*.

MORACEAE
Mulberry Family

Trees and shrubs, rarely herbaceous, with milky latex. Leaves simple, usually alternate, with 2 caducous stipules which leave a scar. Flowers unisexual (the species monoecious or dioecious), small, actinomorphic, borne in heads, disks, or in flat or vase-like "receptacles". Staminate flowers with 4[0] sepals; no petals; 4 stamens, these opposite the sepals. Pistillate flowers with or without a calyx and often with one of the two carpels aborted, ovary superior or inferior. Fruit an achene, nut, drupe, multiple *(Maclura)* or syconium *(Ficus)*.

53/1400; pantropical. *Morus, Maclura,* and *Ficus* are native to the U.S. Economically the family is of some importance, as the source of several food plants and a few ornamentals.

Selected Genera:

Morus (12)- mulberry

Artocarpus (47)- breadfruit, jackfruit; tropical food plants

Broussonetia (7)- paper mulberry, an ornamental

Maclura (12)- *M. pomifera* is the Osage orange or bois d'arc

Ficus (1000)- fig, Bo-tree, banyan tree, Indian rubber tree

Dorstenia (120)- a greenhouse oddity with "open figs"

Recognition Characters:

Trees and shrubs with a milky latex and unisexual flowers.

$$K^4 C^0 A^4 G^0 \quad \text{(staminate flower)}$$
$$K^4 C^0 A^0 \underline{G}^{(2)} \quad \text{(pistillate flower)}$$

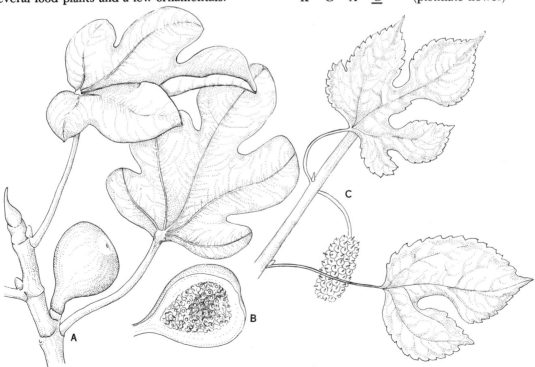

Fig. 31. **Moraceae.** a & b, *Ficus* leaves and syconium in longitudinal section; c, *Morus.*

CANNABACEAE
Hemp Family

Erect annual herbs *(Cannabis)* or climbing perennial herbs *(Humulus)*, with watery juice. Leaves opposite, simple or compound. Flowers unisexual, the species dioecious. Staminate flowers with 5 sepals, no petals, and 5 stamens. Pistillate flowers with 5 sepals, no petals, and a unicarpellate, unilocular gynoecium; ovary superior. Fruit an achene, partially hidden by the persistent calyx.

2/3-5; north temperate zone; widely cultivated. *Humulus* is native to the U.S. The family is the source of hempen fiber, oils, edible seeds, hops, and THC (tetrahydrocannabinols), the psychoactive principles in *Cannabis*. In the older literature the family is often part of the Moraceae. The family spelling used here is conserved by the ICBN over Cannabinaceae, Cannabidaceae, and Cannabiaceae.

Genera:

Cannabis (1-3)- hemp, marijuana, grass, pot (*C. sativa* is the name used by those botanists who consider the genus to be monotypic; others recognize *C. sativa*, *C. indica*, and *C. ruderalis*)

Humulus (2)- *H. lupulus* is the hops of brewing; *H. japonica* is the Japanese hops, an ornamental vine

Recognition Characters:

Erect, aromatic, fibrous herbs with palmately compound leaves; or climbing herbs with opposite leaves and catkin-like pistillate inflorescences of unicarpellate, unilocular flowers.

$$K^5 C^0 A^5 G^0 \quad \text{(staminate flower)}$$
$$K^5 C^0 A^0 \underline{G}^1 \quad \text{(pistillate flower)}$$

URTICACEAE
Nettle Family

Mostly herbs, semishrubs, or rarely small trees with watery juice, sometimes (but **not** always) equipped with stinging hairs; stems usually fibrous. Leaves alternate or opposite, simple, stipulate. Flowers unisexual (the species monoecious or dioecious), small, actinomorphic. Staminate flowers with 4-5 sepals, no petals, 4-5 stamens, opposite the calyx lobes, and often a rudimentary gynoecium. Pistillate flowers with 4-5 sepals [0], no corolla, and a unicarpellate, unilocular gynoecium, the ovary superior or inferior, the placentation basal. Fruit an achene or drupe.

45/550; temperate and tropical regions of both hemispheres. The six genera listed below are native to the U.S. Of little importance economically. Ramie is one of the world's most beautiful and strongest fibers, but it is not widely known. Several species are ornamentals. The young stems and foliage of *Urtica* may be eaten as a potherb.

Selected Genera:

With stinging hairs:

Urtica (35)- nettle, stinging nettle; leaves opposite

Laportea (45)- wood nettle; leaves alternate

Hesperocnide (2)- *H. tenella* occurs in California

Without stinging hairs:

Pilea (200)- clear weed, artillery plant, creeping Charley

Boehmeria (60)- ramie, Chinese silkplant

Parietaria (14)- pellitory

Recognition Characters:

Fibrous herbs with stinging hairs (diagnostic when present!), axillary cymose inflorescences of greenish flowers, these unilocular and unicarpellate.

$$K^{4\text{-}5} C^0 A^{4\text{-}5} G^0 \quad \text{(staminate flower)}$$
$$K^{4\text{-}5} C^0 A^0 \underline{G}^1 \quad \text{(pistillate flower)}$$

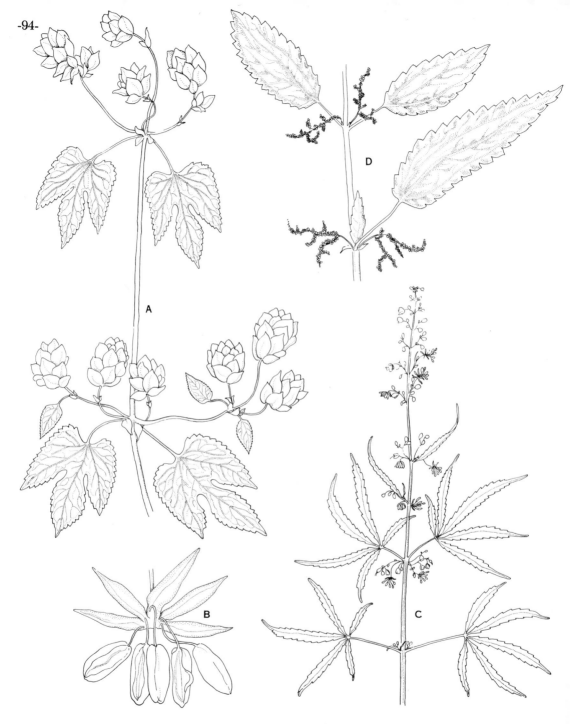

Fig. 32. **Cannabaceae.** a, *Humulus;* b & c, *Cannabis.* **Urticaceae.** d, *Urtica.*

JUGLANDACEAE
Walnut Family

Trees, often resinous and aromatic. Leaves deciduous, alternate, pinnately compound, exstipulate. Flowers unisexual, the species monoecious. The staminate flowers borne in pendulous catkins, the pistillate ones in erect spikes (reduced catkins). Staminate flowers with 3-6[0] sepals, no corolla, 3-40 [100] stamens, and a rudimentary gynoecium. Pistillate flowers with 4 sepals, no petals, and a syncarpous gynoecium of 2 [3] carpels, ovary unilocular by abortion and inferior, style often plumose. Fruit a drupe-like nut or a drupe (depending upon interpretation), enclosed in an involucre ("husk" or "shell").

7/50; north temperate areas and the mountains of the New World tropics. *Carya* and *Juglans* are native to the U.S. Economic products include lumber and edible seeds, and several genera are cultivated as ornamentals.

Selected Genera:

Juglans (40)- walnut; husk indehiscent, pith chambered

Carya (27)- pecan, hickory, bitternut, mockernut; husk dehiscent, pith continuous

Recognition Characters:

Catkin-bearing trees with pinnately compound leaves.

$$K^{3-6} \ C^0 \ A^{3-\infty} \ G^0 \quad \text{(staminate flower)}$$

$$K^4 \ C^0 \ A^0 \ \overline{G}^{(2-3)} \quad \text{(pistillate flower)}$$

Fig. 33. **Juglandaceae.** a & b, *Carya* fruit and branch; c-e, *Juglans* pistillate flower, branch with staminate catkins, and fruit.

FAGACEAE
Beech Family

Trees and shrubs, deciduous or evergreen. Leaves alternate, simple, pinnately veined, variously lobed or entire, stipules deciduous. Flowers typically unisexual (the species usually monoecious). The staminate in catkin-like inflorescences or solitary; the pistillate flowers in an involucre. Staminate flowers of 4-6 [7] sepals, no petals, 4-40 stamens, and a rudimentary gynoecium. Pistillate flowers of 4-6 sepals, no petals, and an epigynous, syncarpous gynoecium of 3 [6] carpels; ovary 3 [6]-loculed, 1-seeded by abortion. Fruit a nut, free or fused to a cup-like organ, the **cupule,** composed of fused bracts. In *Quercus* there is a single fruit/cupule, in *Fagus* 2/cupule, and in *Castanea* 3. The cupule may bear spines.

8/900; of the temperate and tropical regions of the Old World and New World, particularly in the northern hemisphere. *Nothofagus* is southern hemisphere. *Quercus, Lithocarpus, Castanea,* and *Chrysolepis* are native to the U.S. The family is a source of lumber, edible fruits, cork, and many ornamental shade trees.

Selected Genera:

Subfamily: Fagoideae (flowers in axillary clusters)

Fagus (10)- *F. grandifolia* is the eastern beech

Nothofagus (45)- the southern hemisphere beech

Subfamily: Castaneoideae (flowers in catkins; A^{12})

Castanea (12)- *C. dentata* is the American chestnut, its populations much-reduced by the chestnut blight; *C. sativa*, an Old World species, is the source of the edible chestnut of commerce

Chrysolepis (= *Castanopsis*) (2)- *C. chrysophylla* is the chinquapin

Lithocarpus (300) *L. densiflora* is the tan-oak or tanbark-oak

Subfamily: Quercoideae (flowers in catkins; A^6)

Quercus (450)-oak

Subgenus *Leucobalanus*- white oaks

Subgenus *Erythrobalanus*- black oaks and red oaks

Recognition Characters:

Catkin-bearing trees and shrubs with pinnately lobed leaves, buds clustered at the tips of the twigs, and a fruit at least partially enclosed by a cupule; fruit a nut (acorn).

K $^{4-6}$ C ^{0} A $^{4-\infty}$ G ^{0} (staminate flower)

K $^{4-6}$ C ^{0} A ^{0} $\overline{G}$ $^{(3)}$ (pistillate flower)

Fig. 34. **Fagaceae.** a, *Lithocarpus;* b, *Quercus* (white oak group); c, *Quercus* (red oak group); d & e, *Castanea* fruit and branch; f-h, *Fagus* leaves, fruit, and fruit partially enclosed by an involucre.

BETULACEAE
Birch Family

Trees and shrubs. Leaves deciduous, alternate, simple, stipulate, with serrate margins and prominently pinnately veined. Flowers unisexual, the species moneocious. Flowers associated with bracts. Staminate flowers in complex catkins; pistillate flowers in catkin-like spikes. The basic unit of the inflorescence is a cymule, a complex reduced cyme made up of a primary, secondary and tertiary axes, flowers, and primary, secondary, and tertiary bract systems. Various degrees of reduction in this cymule occur in different genera of the family. The staminate flower is composed of 4 sepals in some genera, none in others; no petals; and 2-20 stamens. The pistillate flower lacks a perianth and consist solely of a bicarpellate, syncarpous gynoecium. The ovary is inferior or in some "naked", a term applied when no macroscopic evidence yields information as to ovary position. Fruit a nut or a samara.

6/150; temperate areas of the northern hemisphere. The genera listed below are native to the U.S. The tribes Betuleae and Coryleae are sometimes treated as separate families. When combined, as they are here, the name Betulaceae must be used. The Carpinaceae are also included here. Economic products include lumber, edible seeds, and oil of wintergreen.

Selected Genera:

Tribe: Betuleae (tepals nearly obsolete; nut without an involucre)

Betula (40)- birch; pistillate inflorescence a catkin

Alnus (30)- alder; pistillate inflorescence a woody "cone"

Tribe: Coryleae (tepals present; nut with an involucre)

Carpinus (26)- *C. caroliniana* is the eastern blue beech; involucre flat, 3-lobed

Ostrya (5)- ironwood, hop-hornbeam; involucre sac-like

Corylus (15)- *C. cornuta* is the hazelnut or filbert; involucre large and leaf-like

Recognition Characters:

Catkin-bearing trees with simple, serrate leaves; flowers with a bicarpellate inferior ovary.

$$K \; ^0 \text{ or } ^4 C \; ^0 A \; ^{2\text{-}20} G \; ^0 \quad \text{(staminate flower)}$$

$$K \; ^0 C \; ^0 A \; ^0 \overline{G} \; ^{(2)} \quad \text{(pistillate flower)}$$

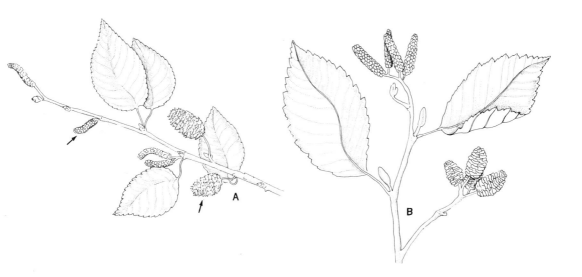

Fig. 35. **Betulaceae.** a, *Betula* (arrows indicate staminate and pistillate catkins); b, *Alnus*.

Magnoliopsida

Order: Family	Common Name	Genera/Species	Distribution	Form	K
Caryophyllales					
Phytolaccaceae	Pokeweed	17/120	Trop. & subtrop.	HST	4-5
Nyctaginaceae	Four o'clock	30/300	Trop.	HST	5
Cactaceae	Cactus	50-220/2000	NW	HS	x
Aizoaceae	Ice plant	130/2500	Trop.	HS	5-8
Molluginaceae	Carpetweed	14/95	Trop. & subtrop.	H	4-5
Caryophyllaceae	Pink	80/2000	Cosmo.	H	5or(5
Portulacaceae	Purslane	19/580	Cosmo.	H	2
Basellaceae	Basella	5/25	Chiefly trop. Am.	H	5
Chenopodiaceae	Goosefoot	102/1500	Cosmo.	HS	5
Amaranthaceae	Pigweed	60-65/900	Temp. & trop.	HS	4-5
Polygonales					
Polygonaceae	Knotweed	40/800	N. temp.	HS	5 or3+
Plumbaginales					
Plumbaginaceae	Leadwort	19/775	Cosmo.	HS	5

aryophyllidae

C	A	G	Fruit Type(s)	Miscellaneous Comments
0	3-∞	1-(x)	ber, drp, utr, ach	Fruits brightly pigmented
0	5	1	ach	Bracts mimic sepals; sepals mimic petals
∞	∞	(2-∞)	ber	Usually spiny succulents
0	∞	(2-5-∞)	cap	Replaces the Cactaceae in the OW
0	5-10	(2-5)	cap	Often weeds of dry places
5[0]	5-10	(2-5)	cap, utr	Free-central placentation; opposite leaves
4-6	4-∞	(2-8)	cap	Often with fleshy leaves
0	5	(3)	ber, drp	Climbing succulents
0	5	(2)	nutl	Several important weeds; halophytes
0	4-5	(2-3)	utr, pyx	Flowers subtended by papery bracts
0	3-9	(3)	ach	Calyx often petaloid
5	5	(5)	utr or cap	*Plumbago, Armeria,* & *Limonium* in U.S.

This subclass consists of four orders, eleven families, and about 11,000 species. Some 90% of the species belong to the Caryophyllales, the most primitive order of the subclass. Plants of the Caryophyllidae are not easily defined on the basis of simple floral or vegetative characters. A high proportion are succulent halophytes. Many have free - central or basal placentation. Betalains, peculiar nitrogenous pigments, are found in and restricted to the Caryophyllales. In some families the perianth consists of a petaloid calyx, in others bracts mimic the calyx and the calyx is petaloid, while a true calyx and corolla are found in some families.

The Caryophyllales (= Centrospermae of older literature) are characterized by free - central or basal placentation and seeds with a curved peripheral embryo which surrounds the perisperm, a storage tissue derived from the nucellus. Nine of the eleven families (excluding the Caryophyllaceae) have betalains rather than anthocyanins. The Polygonales contain the single family Polygonaceae. The family is characterized by its single basal ovule in a syncarpous unilocular gynoecium. It, like the Caryophyllaceae, has anthocyanins.

17/120; tropics and subtropics, particularly in Central American and South Africa. The three genera listed below are native to the U.S. *Phytolacca americana* is a popular potherb, particularly in the South. The species is poisonous, however, and must be prepared properly. A few members of the family are cultivated as ornamentals. As treated here, the Phytolaccaceae includes the Stegnospermataceae, Barbeuiaceae, and the Petiveriaceae.

Selected Genera:

Phytolacca (ca. 35)-pokeweed, pokeberry

Rivina (1)-*R. humilis* is the rouge plant

Petiveria (1)-*P. alliacea* occurs in southern Florida

Recognition Characters:

Herbs with a multicarpellate gynoecium, 1 ovule/locule, and fleshy fruits with highly pigmented juices.

$$K^{4-5}C^0A^{3-\propto}\underline{G}^{1-(x)}$$

PHYTOLACCACEAE
Pokeweed Family

Herbs (ours) or shrubs and trees (tropical); erect or climbing. Leaves alternate, simple, entire, pinnately veined, without stipules. Flowers generally bisexual, actinomorphic, in racemose or cymose inflorescences. Calyx of 4 or 5 sepals. Corolla absent. Androecium of 3-30 stamens, borne on a fleshy disk. Gynoecium of 1-6 free or somewhat fused carpels, stigma number equaling carpel number, 1 ovule per locule, the ovary superior. Fruit a berry, drupe or achene. The embryo circular.

NYCTAGINACEAE
Four O'Clock Family

Herbs (ours), shrubs or trees (tropical). Leaves opposite, simple, entire, exstipulate. Flowers generally bisexual, actinomorphic, often with brightly colored bracts subtending them, these often appearing as a calyx; borne in cymes. Calyx of 5 petaloid sepals. Corolla lacking. Androecium typically of 5 [30] stamens. Gynoecium unicarpellate, unilocular, the ovary superior. Fruit an achene, sometimes enclosed by persistent sepals. The embryo straight or curved.

30/300; tropics, particularly in the New World. Fifteen genera are native to the U.S.

Except as a source of ornamentals, the family is of little economic importance.

Selected Genera:

Mirabilis (60)-Four-O'Clock
Bougainvillea (14)-A widely cultivated vine
Abronia (50)-sand verbena
Oxybaphus (ca. 25)-umbrella wort

Recognition Characters:

Herbs with opposite leaves; bracts mimic the sepals; sepals (sometimes brightly colored) mimic the petals.

$$K^5 C^0 A^5 \underline{G}^1$$

Fig. 36. **Phytolaccaceae.** a, *Phytolacca.* **Nyctaginaceae.** b, *Abronia;* c, *Bougainvillea.*

CACTACEAE
Cactus Family

Succulent herbs and shrubs of diverse form, often spiny. Stems fleshy, spherical, flattened or cylindric, usually assuming the photosynthetic functions of the leaves. Leaves alternate, simple, usually reduced or lacking in mature plants. Plant body often with cushion-like pads called **areoles** from which arise clusters of spines. The areole is usually interpreted as an axillary stem which has failed to develop completely. The spines are reduced leaves borne on the axillary shoot. The areole may also be armed with tiny hooked barbs called **glochidia.** Flowers generally bisexual, actinomorphic, solitary. Calyx petaloid, intergrading into a corolla composed of many petals arranged in several series. The numerous perianth segments may be fused into an elongate hypanthium. Androecium of numerous stamens, free or adnate to the corolla. Gynoecium of 2 to many united carpels, ovary inferior, placentation parietal. Fruit a many-seeded berry, often armed with spines or bristles. Embryo straight or semicircular.

50-220/2000; New World in distribution, occurring from British Columbia to the tip of South America. Cacti are native to all of the conterminous U.S., except Maine, New Hampshire, and Vermont. *Rhipsalis* occurs in Africa, Madagascar, Mauritius, the Seychelles, and Ceylon. Some taxonomists maintain that it is native to these areas. Generic delimitations within the family remain controversial. About 16 genera are native to the U.S. Economic products include handsome ornamentals, edible fruits, and peyote. The spines of some cacti were once used as phonograph needles.

Selected Genera:

Subfamily: Pereskioideae-leaves flat; glochidia absent; no hypanthium

Pereskia (20)-lemon vine

Subfamily: Opuntioideae-leaves small, terete, caducous; glochidia present; no hypanthium

Opuntia (ca. 200)-prickly pear, beavertail, cholla

Subfamily: Cereoideae (=Cacteae)-leaves rudimentary; glochidia absent; hypanthium present

Epiphyllum (20)-old man cactus

Zygocactus (1-2)-*Z. truncatus* is the Christmas cactus

Rhipsalis (60)-mostly epiphytes; perhaps native to Old World

Carnegiea (1)-*C. gigantea* (or *Cereus g.*) is the giant saguaro

Cereus (40)-hedge cactus

Echinocactus (7)-barrel cactus

Echinocereus (70)-hedgehog cactus

Lophophora (3)-*L. williamsii* is peyote, the source of mescaline

Mammillaria (300)-pincushion cactus

Lemaireocereus (25)-*L. thurberi* is the organ-pipe cactus

Ferocactus (35)-barrel cactus

Recognition Characters:

Succulent, often spiny herbs with showy flowers with numerous perianth parts, many stamens, and an inferior ovary.

$$K^X C^\infty A^\infty \overline{G}^{(2-\infty)}$$

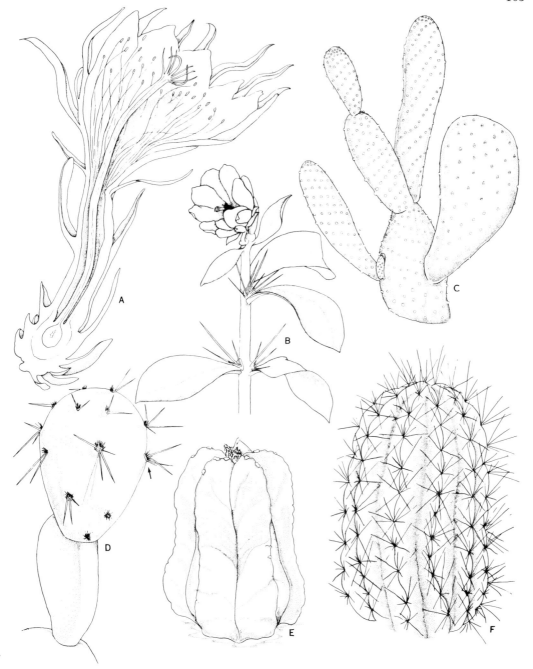

Fig. 37. **Cactaceae.** a, *Epiphyllum* flower in longitudinal section showing hypanthium, tepals, and inferior ovary; b, *Pereskia*, a cactus with well-developed broad leaves at maturity; c, *Opuntia* showing segmented stems; d, *Opuntia* (arrow indicates an areole); e, *Astrophytum*; f, *Echinocactus*.

AIZOACEAE
Ice Plant Family

Xerophytic herbs and low shrubs, often fleshy. Leaves opposite or alternate, simple, fleshy or reduced to minute scales. Flowers bisexual, actinomorphic, solitary or cymose. Calyx of 5-8 sepals. Corolla absent. Androecium of numerous stamens in several series. Gynoecium of 2, 3, 5, or numerous united carpels; placentation axile, sometimes parietal or basal, ovary superior or inferior. Fruit a loculicidal capsule.

130/2500; chiefly South African, with other centers of distribution in Australia, Asia, the western U.S., and South America. Over 100 of the genera have been segregated from *Mesembryanthemum. Sesuvium, Trianthema, Geocarpon, Carpobrotus,* and *Cryophytum* are native to the U.S. Of little economic importance; several genera are cultivated as ornamentals; *Tetragonia* is eaten as a potherb. As treated here, the family includes the Ficoidaceae, Mesembryanthemaceae, and the Tetragoniaceae.

Selected Genera:

Mesembryanthemum (45)-ice plant

Tetragonia (2)-*T. expansa* is the New Zealand spinach

Sesuvium (8)-sea purslane; coastal dunes

Carpobrotus (25)-Hottentot fig; coastal dunes

Lampranthus (200)-widely cultivated ornamentals

Lithops (70)-living rocks, stoneface; ornamentals

Faucaria (36)-tiger jaws; ornamentals

Recognition Characters:

Fleshy herbs, numerous stamens, often with showy petaloid staminodia; the native and introduced species usually coastal.

$$K^{5\text{-}8} C^0 A^\infty \underline{G}(2, 3, 5\text{-}\infty)$$

Fig. 38. **Aizoaceae.** Leaves and flower of *Carpobrotus.*

CARYOPHYLLACEAE
Pink Family

Annual or perennial herbs with swollen nodes. Leaves opposite, simple, entire, often lanceolate, often connected at base by a thin line, usually exstipulate. Flowers bisexual (rarely unisexual, the species dioecious), actinomorphic, solitary or in cymes. Calyx of 5 sepals, free or connate. Corolla of 5 petals, often cleft, rarely lacking. Androecium of 5 or 10 stamens, in 1 or 2 whorls. Gynoecium of 2-5 united carpels, unilocular, with free central placentation; ovules numerous. Fruit a dry capsule, opening by apical teeth (denticidal) or valves; or a utricle. The embryo curved.

80/2000; north temperate and cool regions. About 20 genera are native to the U.S. Of some economic importance as a source of ornamentals; several genera tend to be weedy. As treated here, the family equals the Alsinaceae of older literature and includes the Illecebraceae and Paronychiaceae.

Selected Genera:

Subfamily: Paronychiaeae (corolla absent; calyx aposepalous or synsepalous; fruit a utricle)

Paronychia (40)-whitlow wort; chickweed

Spergula (CA. 5)-spurrey

Spergularia (20+)-sand spurrey

Subfamily: Alsinoideae (corolla present; calyx aposepalous; fruit a capsule)

Sagina (30)-pearlwort

Arenaria (160+)-sandwort

Stellaria (100)-chickweed

Cerastium(100)-mouse-ear chickweed

Subfamily: Silenoideae (corolla present; calyx synsepalous; fruit a capsule)

Gypsophila (126)-baby's breath, an ornamental

Saponaria (30)-S. *officinalis* is the bouncing bet

Dianthus (300)-D. *caryophyllus* is the florist's carnation

Silene (400)-campion, catch-fly

Lychnis (ca. 10)-campion, bladder campion

Agrostemma (1)--A. *githago* is the corn cockle

Recognition Characters:

Herbs with opposite leaves connected by the transverse line; free central placentation; and a dry, many-seeded capsule opening by teeth or valves.

$$K^{5 \text{ or } (5)} C^{5[0]} A^{5-10} \underline{G}^{(2-5)}$$

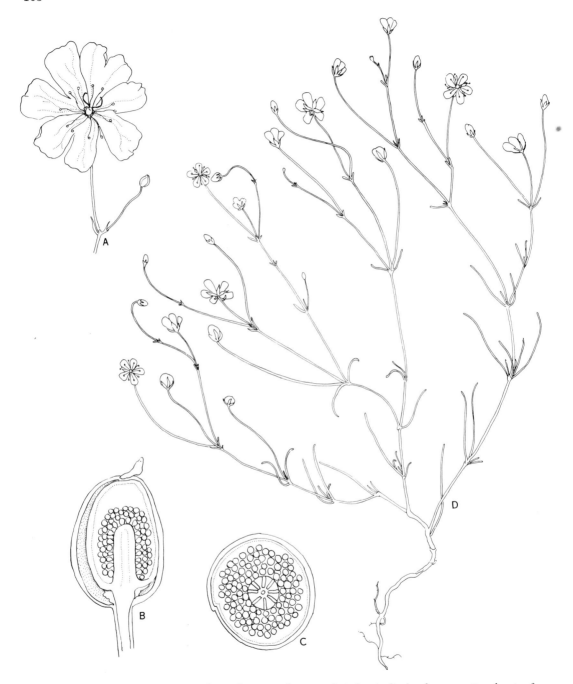

Fig. 39. **Caryophyllaceae.** a, *Cerastium* flower; b & c, *Lychnis* capsule in longitudinal and cross-section showing free central placentation; d, *Arenaria*.

PORTULACACEAE
Purslane Family

Annual or perennial herbs, or subshrubs, often succulent. Leaves alternate or opposite, usually fleshy. Flowers bisexual, actinomorphic, solitary, racemose, or cymose. Calyx of 2 sepals, distinct or connate. Corolla of 4-6 petals (rarely 2 or 3), free or basally connate. Androecium of as many stamens as petals, or twice as many and opposite them, or numerous. Gynoecium of 2-8 united carpels; stigmas and styles usually 2-5; placentation basal; the ovary superior or half-inferior. Fruit a loculicidal capsule, pyxis or rarely a nut.

19/580; cosmopolitan, particularly well represented along the Pacific Coast of the U.S. Nine genera are native to this country. Of little economic importance; several are ornamentals. The floral interpretation adopted here is one of tradition and convenience. It is at odds with more recent work that indicates that the two sepals are actually bracts and that the corolla is a uniseriate perianth of sepals.

Selected Genera:

Portulaca (100+)-purslane; a widespread weed

Lewisia (10)-bitter root

Montia (50+)-miner's lettuce; often merged with *Claytonia*

Claytonia (ca. 20)-spring beauty

Calandrinia (150)-red maids

Recognition Characters:

Herbs with fleshy leaves, 2 sepals, and basal placentation.

$$K^{2 \text{ or } (2)} C^{4-6} A^{4-\infty} \underline{G}^{(2-8)}$$

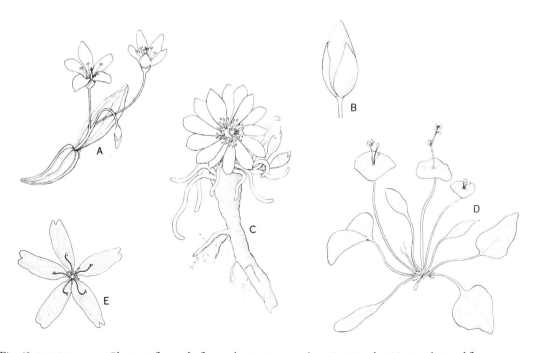

Fig. 40. **Portulacaceae.** a, *Claytonia* flower; b, flower showing two sepals; c, *Lewisia*; d-e, *Montia* plant and flower.

CHENOPODIACEAE
Goosefoot Family

Annual or perennial herbs and shrubs of xerophytic and halophytic habitats. Stems sometimes jointed and essentially leafless. Leaves alternate (opposite in *Salicornia*), simple, sometimes fleshy or reduced to scales, exstipulate. Flowers unisexual (the species monoecious or dioecious) or bisexual, inconspicuous, green, actinomorphic, often bracteate (but **not** scarious), cymose. Calyx of 2-5 connate sepals, usually 5, rarely 0. Corolla absent. Androecium of as many stamens as sepals and opposite them. Gynoecium of 2 (rarely 3-5) united carpels, unilocular, 1-ovuled, the ovary superior (half-inferior in *Beta*). Fruit an indehiscent nutlet. The embryo coiled.

102/1500; cosmopolitan, but characteristically on halophytic soils. Centers of distribution include Australia, the pampas of S. America, the Mediterranean, the Karroo of S. Africa, the shores of the Red Sea, the southwest Caspian coast, central and eastern Asia. Fourteen genera are native to the U.S., mostly in the West. Economic products include several food plants and weeds, such as the Russian thistle.

Selected Genera:

Chenopodium (200+)-goosefoot, lambs-quarters

Atriplex (150)-salt bush

Spinacia (3)-*S. oleracea* is spinach

Beta (12)-*B. vulgaris* is the table beet, sugar beet, and chard

Kochia (80)-summer cypress, burning bush

Salsola (100+)-*S. pestifera* is the Russian thistle

Sarcobatus (2)-*S. vermiculatus* is greasewood

Salicornia (30+)-glasswort, pickleweed, samphire

Halogeton (3)-*H. glomeratus* is halogeton or barilla

Ceratoides (8)-winter fat

Recognition Characters:

Halophytic herbs, often fleshy, with minute green flowers and a unilocular gynoecium with a single ovule. The family is easily confused with the Amaranthaceae from which it differs in lacking dry, papery, sharp-pointed bracts and connate filaments.

$$K^5 C^0 A^5 \underline{G}^{(2)}$$

AMARANTHACEAE
Pigweed Family

Annual or perennial herbs or shrubs. Leaves alternate or opposite, simple, exstipulate. Flowers solitary, in spikes, heads, or racemes, often associated with scarious bracts and bracteoles. Calyx of 4-5 sepals, free or nearly so, minute, dry. Corolla absent. Androecium typically of 4-5 stamens, these opposite the sepals, monadelphous. Gynoecium of 2 or 3 united carpels, unilocular, 1-ovuled (rarely several), ovary superior. Fruit a pyxis or utricle, rarely a drupe or berry. Embryo curved.

60-65/900; temperate and tropical regions of America and Africa. Twelve genera are native to the U.S., particularly well represented in the Southeast. Of little economic importance; a few are ornamentals. *Amaranthus* contains several noxious weeds.

Selected Genera:

Amaranthus (50)-pigweed; a widespread weed

Celosia (60)-cock's comb; an attractive ornamental

Alternanthera (170)-alligator weed; an aquatic

Gomphrena (100+)-globe amaranth; an ornamental

Iresine (70)-blood leaf; an ornamental

Recognition Characters:

Herbs, often with reddish lower stems. Flowers minute, green subtended by papery bracts; the stamens united by their filaments.

$$K^{4-5} C^0 A^{\underline{4-5}} \underline{G}^{(2-3)}$$

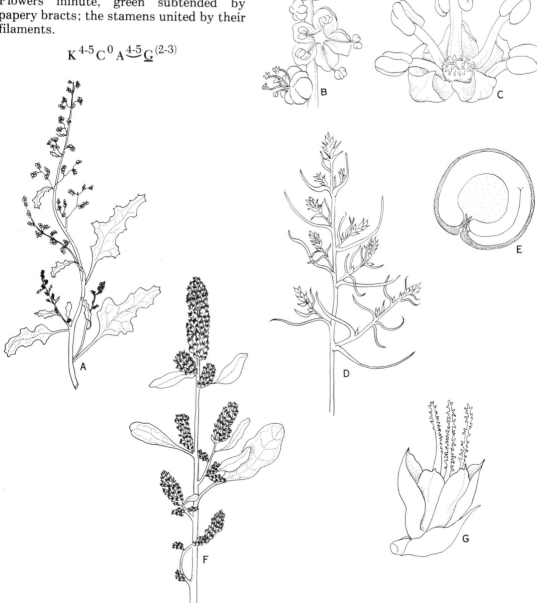

Fig. 41. **Chenopodiaceae.** a-c, *Chenopodium;* d, *Salsola* plant; e, curved embryo. **Amaranthaceae.** f & g, *Amaranthus* plant and flower.

POLYGONACEAE
Knotweed Family

Mostly herbs, sometimes shrubs, rarely trees (tropics). Stems with swollen nodes, hence the family name ("many knees"). Leaves alternate or rarely opposite, often expanded into a membranous sheath (ocrea), forming a collar about the node. Flowers usually bisexual (if unisexual, the species monoecious or dioecious), actinomorphic, typically small, in racemes, spike-like panicles, or heads. Calyx of 3-6 sepals, usually 5 or 3+3 (see below), petaloid. Corolla absent. Androecium of 3-9 stamens, in 2 series. Gynoecium of 3 (rarely 2 or 4) united carpels, unilocular, 1-ovuled, ovary superior. Fruit a lenticular or 3-sided achene. Embryo curved.

40/800; chiefly through the north temperate areas; a few are tropical. Fifteen genera are native to the U.S. The family is best represented in the western states, particularly California, where three genera *(Hollisteria, Gilmania,* and *Nemacaulis)* are endemic. Economic products include food plants and a few ornamentals.

Selected Genera:

A. Flowers cyclic $(K^{3+3}C^0A^{3-9}\underline{G}^{(3)})$

Subfamily: Rumicoideae

Tribe: Eriogoneae (ocrea lacking)

Eriogonum (ca. 200)-false buckwheat; mostly West

Tribe: Rumiceae (ocrea present)

Rumex (200)-dock, sorrel; widespread weeds

Rheum (40)-*R. rhaponticum* is rhubarb

Oxyria (1)-*O. digyna* is the mountain sorrel

A. Flowers acyclic $(K^5C^0A^{5-8}\underline{G}^{(3)})$

Subfamily: Polygonoideae (endosperm not mottled)

Tribe: Polygoneae

Polygonum (ca. 200)-knotweed, smartweed

Fagopyrum (2)-*F. esculentum* is the buckwheat

Subfamily: Coccoloboideae (endosperm mottled)

Tribe: Coccolobeae

Muehlenbeckia (20)-tapeworm plant; cladophylls

Coccoloba (ca. 125)-sea grape; tropical

Antigonon (3-4)-mountain rose vine; ornamental

Recognition Characters:

Herbs with swollen nodes, ocreae or involucrate heads, petaloid calyx, and a lens-shaped or triangular achene, often black. Beginning students are often confused by an apparent 3-3-6-3 floral pattern and attempt to place members of this family in the Liliaceae. Beware. Note the two floral patterns:

$$K^{3+3}C^0A^{3-9}\underline{G}^{(3)} \quad \text{or} \quad K^5C^0A^{5-8}\underline{G}^{(3)}$$

Fig. 42. **Polygonaceae.** a & b, *Rumex* in fruiting phase and plant; c, *Cocoloba* (arrow indicates ocrea); d - f, *Eriogonum* inflorescence, plant, and flower.

Magnoliopsida

Order: Family	Common Name	Genera/Species	Distribution	Form	K
Dilleniales					
Paeoniaceae	Paeony	1/33	N. temp.	H	5
Crossosomataceae	Crossosoma	1/4	N. Am.	S	(5)
Theales					
Theaceae	Tea	16/500	Trop. & subtrop.	TS	5
Elatinaceae	Water wort	2/40	Trop. & temp.	SH	3-5
Guttiferae	Garcinia	40/1000	Trop.	TS	4-5
Malvales					
Tiliaceae	Basswood	50/450	Trop. & temp.	TSH	5
Sterculiaceae	Sterculia	60/700	Trop. & temp.	TSH	(3-5)
Bombacaceae	Bombax	20/180	Trop.	T	(5)
Malvaceae	Mallow	85/1500	Cosmo., esp. trop.	HST	3-5
Sarraceniales					
Sarraceniaceae	Pitcher-plant	3/17	NW only	H	4-5
Droseraceae	Sundew	4/105	Plants of acid bogs	H	(5)
Violales					
Violaceae	Violet	22/900	Cosmo.	HS	5
Turneraceae	Turnera	7/120	Trop.	TSH	(5)
Passifloraceae	Passion flower	12/600	Trop. & temp.	VST	3-5
Bixaceae		3/30	Trop.	TS	4-5
Cistaceae	Rock-rose	8/200	N. temp.	SH	5
Tamaricaceae	Tamarisk	4/120	Temp. & subtrop.	ST	(4-5)
Frankeniaceae	Frankenia	4/90	Trop. & temp.	H	(4-7)
Fouquieriaceae	Ocotillo	2/11	N. Am.	ST	5
Caricaceae	Papaya	4/55	Trop. Am. & Afr.	T	5
Loasaceae	Loasa	15/250	Trop. & subtrop.	H	5

Dilleniidae

C	A	G	Fruit Type(s)	Miscellaneous Comments
5-10	∞	<u>2-5</u>	fol	Common in cultivation
5	∞	<u>3-6</u>	fol	Endemic to Calif. & Ariz.
5	∞	(2-10)	cap, drp	*Camellia sinensis* is tea
3-5	3-10	(3-5)	cap	Freshwater sites
4-5	∞	(3-5)	cap, ber, drp	Leaves punctate
5	∞	(2-10)	cap, schizo	*Tilia* common in U.S.
5	5-x	(5)	schizo	Family of cacao and cola nut
5	5-∞	(2-5)	cap	None is native to U.S.
5	∞	(5-∞)	cap, schizo	Often with stellate pubescence
5	x-∞	(5-6)	cap	Insectivorous bog plants
5	5	(2-5)	cap	Sundew and Venus flytrap native
5z	5	(3)	cap, ber	Flower often spurred
5	5	(3)	cap	*Turnera* native to Texas
3-5	3-5	(3)	cap, ber	Vines with quite elaborate flowers
4-5	∞	(2-5)	cap	Includes the Cochlospermaceae
5	∞	(3,5-10)	cap	Four genera native to U.S.
4-5	8-10	(4-5)	cap	Gymnosperm-like foliage
4-7	6	(3)	cap	Saline and dry sites
(5)	10-15	(3)	cap	*Fouquiera* and *Idria* are desert plants
(5)	10	(5)	ber	Papaya plants are unisexual
5	5-∞	(3-7)	cap	Often with stinging hairs

Begoniaceae	Begonia	5/920	Trop.	H	2-5
Datiscaceae	Datisca	3/4	OW & NW	HT	3-9
Cucurbitaceae	Squash	100/850	Trop.	V	5
Salicales					
Salicaceae	Willow	2-3/350-500+	Cosmo.	TS	o-x
Capparales					
Capparaceae	Caper	46/800	Trop. & Subtrop.	TS	4
Cruciferae	Mustard	375/3200	N. temp.	HS	4
Resedaceae	Mignonette	6/70	Chiefly Medit.	HS	2-8
Moringaceae	Moringa	1/12	Afr. to Ind.	T	5
Ericales					
Cyrillaceae	Cyrilla	3/13	NW	ST	5
Clethraceae	Pepperbush	1/30	OW & NW	ST	5
Ericaceae	Heath	50-82/2500	Cosmo; acid soils	ST	4-5
Empetraceae	Crowberry	3/10	NW	S	3
Pyrolaceae	Pyrola	4/46	N. temp.	H	4-5
Monotropaceae	Indian pipe	12/30	Temp; OW & NW	H	2-6
Diapensiales					
Diapensiaceae	Diapensia	6/20	N. hemis.	SH	5 or (5
Ebenales					
Sapotaceae	Sapote	35-75/800	Trop.	TS	4-10
Ebenaceae	Ebony	4/450	Trop.	TS	3-7
Styracaceae	Storax	12/180	NW; Medit., As.	ST	(5-4)
Symplocaceae	Symplocos	2/500	OW & NW trop.	TS	5
Primulales					
Myrsinaceae	Myrsine	35/1000	Trop. & Subtrop.	TS	4-6
Primulaceae	Primrose	20/1000	Cosmo.	H	5

0	∞	$\overline{(2\text{-}3)}$	cap	None is native to the U.S.
0	8-∞	$\overline{(3\text{-}5)}$	cap	California only in U.S.
(5)	$\overset{\frown}{5}$	$\overline{(3)}$	pepo	Typically vines with unisexual flowers
0	2-x	$\underline{(2)}$	cap	Seeds comose
4z	4-∞	$\underline{(2)}$	cap, ber	Ovary stipitate
4	4 + 2	$\underline{(2)}$	sil, slq	Often with acrid taste
0-8	3-∞	$\underline{(2\text{-}7)}$	cap, ber	*Oligomeris* native to the U.S.
5	5 + 5	$\underline{(3)}$	cap	Cultivated in Fla. and Calif.
5	5 or 10	$\underline{(2\text{-}4)}$	cap, ber	*Cyrilla & Cliftonia* endemic to SE U.S.
5	5 + 5	$\underline{(3)}$	cap	In SE U.S.
(4-5)	8-10	$\underline{(4\text{-}5)}$	cap, drp, ber	Usually on acid soils
3	3	$\underline{(2\text{-}9)}$	drp or ber	Heath-like shrubs
4-5	10	$\underline{(4\text{-}5)}$	cap	Autotrophic; anthers with terminal pores
3-6	6-12	$\underline{(4\text{-}6)}$	cap	Parasitic; anthers opening by longitudinal slits
(5)	5	$\underline{(3)}$	cap	In eastern U.S.
4-10	8-15	$\underline{(4\text{-}5)}$	ber	Important source of fruits and gums
3-7	=1-3C	$\overline{(2\text{-}16)}$	ber	Persimmon tree is only native species
(5-4)	=1-2C	$\overline{(3\text{-}5)}$	drp	*Halesia* and *Styrax* in U.S.
(5-10)	5-∞	$\overline{(2\text{-}5)}$	ber, drp-like	*Symplocos tinctoria* in U.S.
(4-6)	4-6	$\overline{(4\text{-}6)}$	drp, ber	*Rapanea* and *Ardisia* native to Florida
(5)	5	$\underline{(5)}$	cap, pyx	Eleven genera native to U.S., mostly in E

The Dilleniidae consist of twelve orders, 69 families, and almost 24,000 species. The subclass is coordinate with the Rosidae, both being more advanced than the Magnoliidae, but less so than the Asteridae. More than one-third of the species have parietal placentation; about one-third are sympetalous. Members of the subclass differ from those of the Magnoliidae in being syncarpous, except for the Dilleniales which appears to be the link between the two.

The orders are characterized as follows. The Dilleniales are clearly the most primitive order, differing from the Magnoliidae in the sequence of stamen initiation. The Theales are mostly woody plants with numerous stamens. The Malvales usually have numerous stamens, often united by their filaments. The Sarraceniales are herbs with simple leaves modified for trapping insects. The Violales are difficult to define, but most have a unilocular ovary with parietal placentation. The Salicales consist of the Salicaceae, the only amentiferous family of the subclass. The Capparales are generally defined by the characters of the Cruciferae and Capparales, with their 4-merous flowers. The Ericales are typically sympetalous with axile placentation, numerous ovules per locule, and appendaged anthers. The Ebenales are woody plants with sympetalous corollas, axile placentation and usually only one or two ovules per locule. In the Primulales the fertile stamens are opposite the corolla lobes.

GUTTIFERAE (=CLUSIACEAE)
Garcinia Family

Trees, shrubs, and herbs. Leaves opposite, simple, entire, exstipulate. Oil glands or passages present, often appearing as translucent dots on the leaves. Flowers unisexual (Clusioideae) or bisexual (Hypericoideae), in cymes, frequently umbellate. Calyx of 4-5 separate sepals. Corolla of as many separate petals as sepals. Androecium of numerous fascicled stamens. Gynoecium of 3 or 5 united carpels, usually with 3-5 locules, placentation axile, ovules numerous, ovary superior. Fruit a capsule, sometimes a berry or drupe.

40/1000; temperate and tropical regions. *Hypericum, Ascyrum, Triadenum,* and *Clusia* are native to the U.S. There is some confusion as to the proper name for this family. The Clusiaceae is the correct alternative if you prefer family names ending in -aceae. The Hypericaceae is not a proper alternative for either the Guttiferae or Clusiaceae. It is a subfamily of the Guttiferae which is often recognized as a distinct family. Of little economic importance in North America. A few are ornamentals.

Selected Genera:

Subfamily: Clusioideae (flowers unisexual; = Clusiaceae s.s.)

Clusia (200)-autograph tree; native to southern Florida

Garcinia (220)-*G. mangostana* is the mangosteen, one of most prized tropical fruits

Subfamily: Hypericoideae (flowers bisexual; = Hypericaceae s.s.)

Hypericum (350)-St. John's wort

Ascyrum (5)-St. Andrew's cross

Recognition Characters:

Herbs or shrubs with opposite, dotted leaves and often yellow flowers with numerous stamens in fascicles.

$$K^{4-5} C^{4-5} A^{\infty} \underline{G}^{(3) \text{ or } (5)}$$

Fig. 43. **Guttiferae.** *Hypericum.* a, leaves and flower (arrow indicates punctations in a leaf blade); b, stamen bundle.

50/450; of the tropical and temperate regions, particularly southeast Asia and Brazil. Three genera *(Tilia, Triumfetta,* and *Corchorus)* are indigenous to the U.S. Economic products include lumber, ornamentals and fibers.

Selected Genera:

Tilia (ca. 50)-basswood, linden tree

Corchorus (40)-jute, a widely used fiber

Sparmannia (7)-a popular ornamental

Recognition Characters:

In the U.S. the predominant genus is *Tilia,* a genus of trees with cordate leaves and cymose flowers borne on a subtending leaflike bract.

$$K^5 \, C^5 \, A^\infty \, \underline{G}^{(2\text{-}10)}$$

TILIACEAE
Basswood Family

Trees or shrubs, rarely herbs; often with stellate pubescence. Leaves alternate, simple, stipulate. Flowers bisexual, actinomorphic, cymose. Calyx of 5 [3 or 4] separate or basally connate sepals. Petals as many as sepals, or absent. Androecium of numerous stamens. Gynoecium of 2-10 united carpels and as many locules, ovary superior. Fruit a capsule or schizocarp.

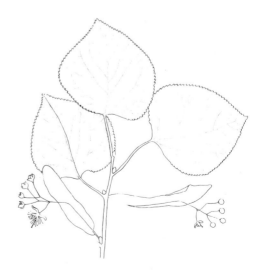

Fig. 44. **Tiliaceae.** Leaves, flowers, and fruits of *Tilia.*

MALVACEAE
Mallow Family

Herbs, shrubs, or rarely small trees, often covered with stellate hairs. Leaves alternate, simple, often palmately lobed and veined, stipulate. Flowers usually bisexual, actinomorphic, solitary or in cymes. Calyx of 3-5 sepals, ± united, sometimes subtended by coherent bracts (the **epicalyx**). Corolla of 5 separate petals. Androecium of numerous stamens, monadelphous. Gynoecium of 5 to many united carpels, rarely unicarpellate, ovary superior. Fruit a capsule, schizocarp, sometimes a berry or samara.

85/1500; cosmopolitan, particularly well represented in the American tropics. Twenty-seven genera are native to the U.S. Of moderate economic importance because of cotton fibers derived from the seeds, several ornamentals, and a few food plants.

Selected Genera:

Tribe: Malveae (fruit a schizocarp)

Abutilon (150)-flowering maple, velvet leaf

Sphaeralcea (60+)-globe mallow

Malvastrum (12)-false mallow

Lavatera (25)-tree mallow, malva rosa

Althaea (75)-hollyhock, marsh mallow

Malva (30)-mallow, cheeseweed

Sida (180+)-alkali mallow

Anoda (15)-snowcup, a popular ornamental

Sidalcea (25)-checkers

Tribe: Hibisceae (fruit a capsule)

Hibiscus (200)-hibiscus; *H. esculentus* is okra

Gossypium (20-67)-*G. hirsutum* is upland cotton of the South

Recognition Characters:

Herbs and shrubs with stellate pubescence, palmately lobed and veined leaves, numerous stamens united by their filaments, and many-carpellate fruits, coherent or splitting into wedges.

$$K^{\underline{3\text{-}5}} C^5 A^{\underline{\infty}} \underline{G}^{(5\text{-}8)}$$

Fig. 45. **Malvaceae.** a & b, *Malva* fruits (arrow indicates epicalyx); c, *Hibiscus;* d, *Hibiscus* flower in longitudinal section (arrow indicates staminal column); e, dehisced capsule.

SARRACENIACEAE
Pitcher Plant Family

Insectivorous perennial herbs. Leaves tubular, with reduced blades. Flowers bisexual, actinomorphic, large, nodding, solitary or in racemes, subtended by 1-3 bracteoles. Calyx of 4-5 separate sepals, these often colored. Corolla of 5 free petals (rarely 0). Androecium of numerous stamens, sometimes 12-15. Gynoecium of 5-6 [3] united carpels, ovary superior, style simple and dilated into an umbrella-like structure. Fruit a loculicidal capsule.

3/17; restricted to the New World. *Sarracenia* occurs from Labrador to the southeastern coastal plain of the U.S., and westward to Texas. *Darlingtonia* is found in southern Oregon and northern California. The family is of no economic importance.

Genera:

Sarracenia (9)-pitcher plant, trumpets

Darlingtonia (1)-*D. californica* is the cobra lily

Heliamphora (6)-restricted to Venezuela and Guyana

Recognition Characters:

Bog plants with tubular or vase-like leaves (hooded in the cobra lily) filled with dead insects.

$$K^{4-5} C^5 A^{x-\infty} \underline{G}^{(5-6)}$$

DROSERACEAE
Sundew Family

Perennial, insectivorous, glandular herbs, often of acid bogs. Leaves either covered with sticky glands which trap insects or sometimes equipped with sensitive hairs which trigger a folding mechanism when touched. Flowers bisexual, actinomorphic. Calyx of 5 fused sepals. Corolla of 5 separate petals. Stamens 5. Gynoecium of 2, 3, or 5 united carpels, the ovary superior. Fruit a loculicidal capsule.

4/105; *Drosera* is cosmopolitan; the other genera are monotypic and restricted in their distribution. *Drosera* and *Dionaea* are native to the U.S. Of no economic importance; several are cultivated as novelties because of their insectivorous habit. As treated here, the family includes the Dionaeaceae.

Genera:

Drosera (90)-sundew

Drosophyllum (1)-*D. lusitanica;* western Mediterranean

Dionaea (1)-*D. muscipula* is the Venus fly trap; endemic to North and South Carolina

Aldrovanda (1)-*A. vesiculosa*, a submerged aquatic with leaves like those of *Dionaea;* much of the Old World

Recognition Characters:

Insectivorous plants, the leaves either covered with sticky glandular hairs or equipped with sensitive hairs which close the blade to form a trap.

$$K^{(5)} C^5 A^5 \underline{G}^{(2-3-5)}$$

Fig. 46. **Droseraceae.** a & b, *Dionaea* plant and flower; c & d, *Drosera* plant and flower. **Sarraceniaceae.** e & f, *Darlingtonia* plant and flower; g & h, *Sarracenia* flower and plant.

VIOLACEAE
Violet Family

Perennial herbs (rarely annual) or shrubs. Leaves alternate, simple, sometimes variously lobed or dissected, stipulate. Flowers bisexual, zygomorphic or actinomorphic, sometimes cleistogamous. Calyx of $5 \pm$ separate sepals. Corolla of 5 unequal petals, the lower often enlarged and spurred. Androecium of 5 stamens, connivent about the ovary, the abaxial one often spurred. Gynoecium of 3 united carpels, unilocular, superior, placentation parietal. Fruit an explosive loculicidal capsule or berry.

16/850 species, cosmopolitan. Two genera, *Viola* and *Hybanthus,* are native to the U.S. Of no real economic importance other than as a source of ornamentals; many species of *Viola* are cultivated.

Selected Genera:

Viola (450)-violet

Rinorea (280)-a tropical tree with regular flowers

Hybanthus (80)-green violet

Recognition Characters:

Herbs (ours) with a 5-merous, zygomorphic corolla, 3-carpellate gynoecium, spurred flower, spurred anthers, and an explosive capsule.

$$K^5 \, Cz^5 \, A^5 \, \underline{G}^{(3)}$$

Fig. 47. **Violaceae.** a, plant; b, flower (arrow indicates spur); c, flower in section (arrow indicates appendaged stamens); d, capsule. All drawings show *Viola.*

PASSIFLORACEAE
Passion Flower Family

Herbaceous tendril-bearing climbers (ours) or shrubs and trees (tropical). Leaves alternate, entire or lobed, often with glands on the petiole. Flowers usually bisexual, actinomorphic, axillary. Calyx of 3-5 free or partially connate sepals. Corolla of 3-5 petals. Corona of 1 or more series of thread-like filaments. Androecium of 3-5 [10] stamens. Gynoecium of 3 united carpels, unilocular, sometimes on a stalk, placentation parietal. Fruit a berry or loculicidal capsule.

12/600, warm temperate and tropical. The center of distribution is tropical America. *Passiflora* is indigenous to the U.S. A source of several ornamentals and edible fruits. The common name of the family is a reference to the passion of Christ, not the more commonly encountered variety. Various parts of the flower have symbolic meaning.

Selected Genus:

Passiflora (400)-passion flower; passion fruit and granadilla, both highly prized tropical fruits.

Recognition Characters:

Climbing herbs (ours) with conspicuous flowers with fringed corona and stalked gynoecium.

$$K^{3\text{-}5} C^{3\text{-}5} A^{3\text{-}5} \underline{G}^{(3)}$$

Fig. 48. **Passifloraceae.** Leaf and flowers of *Passiflora*.

CUCURBITACEAE
Gourd Family

Annual or perennial herbs, climbing or prostrate, often scabrid, usually with spirally coiled tendrils. Leaves alternate, simple, often palmately 5-lobed and palmately veined. Flowers generally unisexual (the species monoecious or dioecious), actinomorphic, large, axillary. Staminate flowers with 5 sepals; 5 petals, usually sympetalous; 5 stamens, although usually highly modified by cohesion and twisting, etc. Pistillate flower with 5 sepals, 5 petals, and a tricarpellate inferior ovary. The gynoecium has a single locule and parietal placentation. Fruit a berry (often referred to as a pepo) with a leathery or hard exocarp.

100/850; throughout the warmer regions of the Old and New World. About 14 genera are native to the U.S. The family is economically important as a source of many food plants and ornamentals. The classification used below follows Jeffreys, C. 1972 (Kew Bulletin 15:337-371).

Selected Genera:

Tribe: Joliffieae

    Momordica (60)-balsam pear, balsam apple

Tribe: Benincaseae

    Citrullus (4)-*C. vulgaris* is the watermelon

    Luffa (8)-*L. cylindrica* is the vegetable sponge

    Ecballium (1)-*E. elaterium* is the squirting cucumber

    Lagenaria (1)-*L. vulgaris* is the bottle gourd

Tribe: Cyclanthereae

    Echinocystis (25)-wild cucumber

    Marah (7)-wild cucumber, man root

Tribe: Sicyoeae

    Sechium (1)-*S. edule* is chayote or yuca, a peculiar 1-seeded squash

Tribe: Cucurbiteae

    Cucurbita (20)-pumpkin, squash, gourd

Tribe: Melothrieae

    Cucumis (40)-muskmelon, canteloupe, gherkin, cucumber

Recognition Characters:

Coarse, tendril-bearing vines with unisexual flowers (often yellow) with inferior ovaries and leathery berries (pepo).

$$K^5 \, C^{(5)} A^5 \, G^0 \quad \text{(staminate flower)}$$

$$K^5 \, C^{(5)} A^0 \, \overline{G}^{(3)} \text{(pistillate flower)}$$

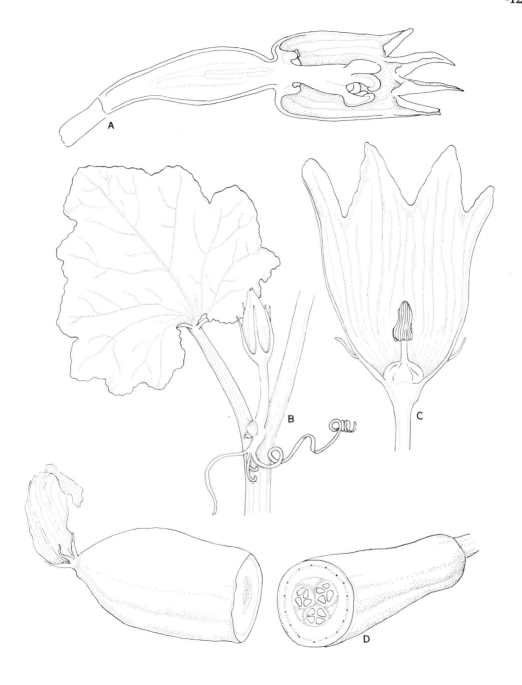

Fig. 49. **Cucurbitaceae.** a, pistillate flower; b, leaf, flower, and tendril; c, staminate flower; d, pepo in x-s.

SALICACEAE
Willow Family

Trees and shrubs. Leaves alternate, simple, stipulate. Flowers unisexual, the species dioecious. Staminate and pistillate flowers in separate erect or pendulous catkins, these often appearing before the leaves. Flowers "subtended" by fringed or hairy bracts and a cup-like disk or 1 or 2 glands; the disk or gland(s) possibly representing the calyx. Corolla absent. Staminate flowers of 2 or more stamens and subtending bracts and glands. Pistillate flowers of 2-4 united carpels, unilocular, ovary superior, stigmas 2-4. Fruit a capsule. Seeds comose.

2-3/350-500+; cosmopolitan. *Salix* and *Populus* are native to the U.S. Of little economic importance, except as a source of ornamentals.

Genera:

Salix (300-500)-willow; buds with a single bud scale

Populus (40)-popular, quaking aspen; buds with several bud scales, the petioles often flattened

Chosenia (1)-a willow-like Asian tree

Recognition Characters:

Catkin-bearing dioecious trees and shrubs with comose seeds.

$$K^x \ C^0 \ A^{2-x} \ G^0 \text{ (staminate flower)}$$

$$K^0 \ C^0 \ A^0 \ \underline{G}^{(2)}\text{(pistillate flower)}$$

Fig. 50. **Salicaceae.** *Populus.* a, staminate catkin (note multiple bud scales); b & c, fruit and seed; d, fruiting branch and leaves. *Salix.* e & f, fruits; g, pistillate flower; h, staminate catkins; i, staminate flower; j, branch with leaves; k, buds (note single bud scale).

CAPPARACEAE
Caper Family

Trees, shrubs, rarely herbs, sometimes climbing. Leaves alternate, simple or palmately compound, stipules rudimentary or lacking. Flowers generally bisexual, usually zygomorphic. Calyx of 4 separate sepals. Corolla of 4 separate petals. Androecium of 4 to many stamens. Gynoecium of 2 united carpels, the placentation parietal, the ovary unilocular and superior, usually stipitate. Fruit a capsule or berry.

46/800; tropical and subtropical areas of the world. Many are xerophytes. Eight or nine genera are native to the U.S. Of economic importance as a source of ornamentals and capers, a salad seasoning. The spelling of the family name was conserved by the ICBN over the more widely used Capparidaceae. As treated here, the family includes the Cleomaceae.

Selected Genera:

Capparis (350)-the flower buds of *C. spinosa* are capers

Cleome (including *Polanisia*) (ca. 200)-spider plant

Wislizenia (10)-jackass clover; western

Cleomella (20)-stinkweed

Isomeris (1)-*I. arborea* is the bladderpod

Recognition Characters:

(Ours) herbs or shrubs with zygomorphic flowers, and a stipitate, unilocular ovary with parietal placentation. The family is most easily confused with the Cruciferae.

$$K^4 \; Cz^4 \; A^{4-\infty} \; \underline{G}^{(2)}$$

Fig. 51. **Capparaceae.** a & b, plant and flower detail of *Cleome*.

Table 8. Comparison of the Capparaceae and Cruciferae.

Character	Capparaceae	Cruciferae
symmetry	zygomorphic	actinomorphic
stamens	4-many	4 + 2
locules	1	2
ovary	stipitate	rarely stipitate
fruit	capsule	silique

CRUCIFERAE (=BRASSICACEAE)
Mustard Family

Annual to perennial herbs, a few somewhat woody, with watery, acrid juices. Leaves usually alternate, simple, exstipulate, with simple, forked or stellate hairs. Flowers bisexual, actinomorphic, usually racemose. Calyx of 4 separate sepals. Corolla of 4 separate petals, these often clawed. Androecium of 6 stamens, tetradynamous in a 4 long and 2 short configuration. Gynoecium of 2 united carpels (several authors maintain that there are 4 carpels) with parietal placentae, often divided by a false septum, ovary superior. Fruit a dehiscent silique (if 3 or more times longer than wide) or a silicle (if short and squatty); rarely indehiscent or transversely dehiscent. The nature of the mature fruit is of utmost importance in identification of plants in this family.

375/3200; widely distributed through the cooler regions of the northern hemisphere, rare in the tropics. About 55 genera are native to the U.S., particularly in the West. Many food plants, condiments, ornamentals, and weeds make the family important economically.

Selected Genera:

Tribe: Stanleyeae

Stanleya (6)-prince's plume; ovary stipitate

Tribe: Streptantheae

Streptanthus (8)-stamens often in 3 pairs

Tribe: Sisymbrieae

Sisymbrium (80)-tumble mustard

Descurainia (46)-tansy mustard; toxic to cattle

Isatis (30)-*I. tinctoria* is woad, an ancient dye plant

Tribe: Hesperidae

Erysimum (80)-wall flower

Hesperis (24)-dame's violet, rocket

Tribe: Arabideae

Barbarea (12)-winter cress, golden rockets

Armoracia (3)-*A. rusticana* is the horseradish

Nasturtium (50)-water cress, an edible aquatic

Cardamine (160)-bitter cress

Dentaria (20)-toothwort; often merged with *Cardamine*

Arabis (120)-rock cress

Tribe: Alysseae

Lunaria (3)-honesty, money wort, moon wort; ornamental

Alyssum (100)-madwort, golden tuft

Draba (270)-whitlow grass

Tribe: Lepidieae

Capsella (5)-*C. bursa-pastoris* is the shepherd's purse

Thlaspi (60)-fanweed, a widespread weed

Lepidium (130)-pepper grass, also weedy

Iberis (30)-candytuft, a popular ornamental

Tribe: Brassiceae

Brassica (50)-mustard, cabbage, kale, brussel sprouts, broccoli, cauliflower, rutabaga, kohlrabi; many edible plants derived from *B. oleracea*

Cakile (4)-sea rocket, coastal dunes

Raphanus (8)-*R. sativus* is the radish

Recognition Characters:

Herbs with an acrid taste; flowers of 4 sepals, 4 petals, 6 stamens (4+2), and a silique or silicle.

$$K^4 \; C^4 \; A^{4+2} \; \underline{G}^{(2)}$$

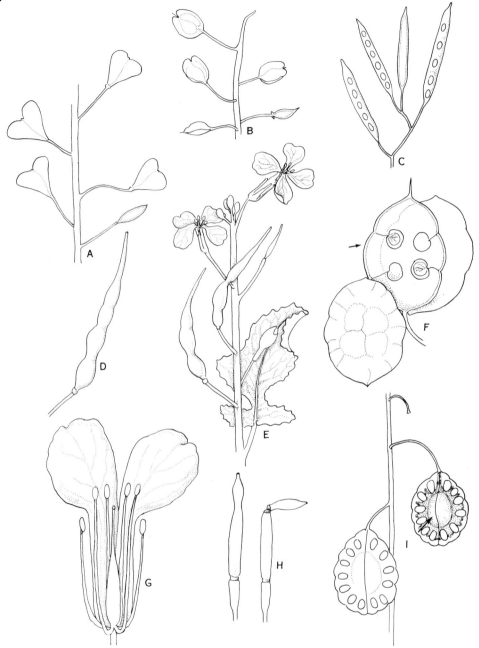

Fig. 52. **Cruciferae.** a, silicles of *Capsella;* b, silicles of *Thlaspi;* c, generalized silique; d, silique of *Raphanus; e, Raphanus;* f, *Lunaria* showing dehisced silicle (arrow indicates the replum); g, generalized flower showing tetradynamous androecium; h, silique dehiscing transversely, unlike most mustards which dehisce longitudinally; i, 1-seeded silicle of *Thysanocarpus* (arrow indicates seed position).

ERICACEAE
Heath Family

Shrubs, subshrubs, somewhat woody perennial herbs, rarely trees and vines. Leaves mostly alternate, simple, often leathery and evergreen, exstipulate. Flowers bisexual, typically actinomorphic; solitary or variously racemose. Calyx of 4-5± connate sepals. Corolla of 4-5 fused petals, often urceolate or campanulate. Androecium of twice as many stamens as petals, or less often the same number, inserted on a disk, the anthers often tailed, opening by terminal pores (rarely by longitudinal slits), pollen in tetrads. Gynoecium of 4-5 united carpels, with as many locules, ovary superior or inferior, placentation axile. Fruit a capsule, drupe, or berry.

50-82/1350-2500 generally distributed on acid soils of the temperate regions of both the northern and southern hemisphere. The Ericaceae are almost absent from Australasia. About 25 genera are indigenous to the U.S. Useful economic products include food plants, oil of wintergreen, and numerous ornamentals. The Ericaceae in the traditional sense included parasitic and hemiparasitic herbs in addition to the autotrophic shrubs. It is particularly among such forms that mycorrhizal relationships are common. As treated here, the family excludes these groups which are now recognized as separate families, the Pyrolaceae and the Monotropaceae. The Vacciniaceae of some authors are, however, retained in the Ericaceae.

Selected Genera:

A. Winter buds with scales; corolla quickly falling

Subfamily: Rhododendroideae (stamens usually without appendages; ovary superior; fruit a septicidal capsule; seeds usually winged)

Phyllodoce (10)- mountain heather

Ledum (5)-Labrador tea

Rhododendron (including *Azalea*) (1300)-rhododendron, azalea, rose bay

Kalmia (6)-mountain laurel

Subfamily: Vaccinioideae (stamens usually appendaged; ovary superior or inferior; fruit a loculicidal capsule, drupe or berry; seeds wingless)

Vaccinium (200)-blueberry-huckleberry- cranberry

Gaultheria (150)-wintergreen, salal

Oxydendrum (1)-*O. arboreum* is sourwood

Arbutus (20)-madrone

Arctostaphylos (70)-manzanita, bearberry

A. Winter buds absent; corolla usually persistent after flowering

Subfamily: Ericoideae

Erica (530)-heath; many ornamental species

Calluna (1)-*C. vulgaris* is heather

Cassiope (12)-white heather

Recognition Characters:

Woody plants, usually shrubs, with leathery, often evergreen leaves, urceolate or campanulate flowers with anthers opening by terminal pores.

$$K \underset{\smile}{4\text{-}5} \; C^{(4\text{-}5)} \; A^{8\text{-}10} \; \overline{\underline{G}}^{(4\text{-}5)}$$

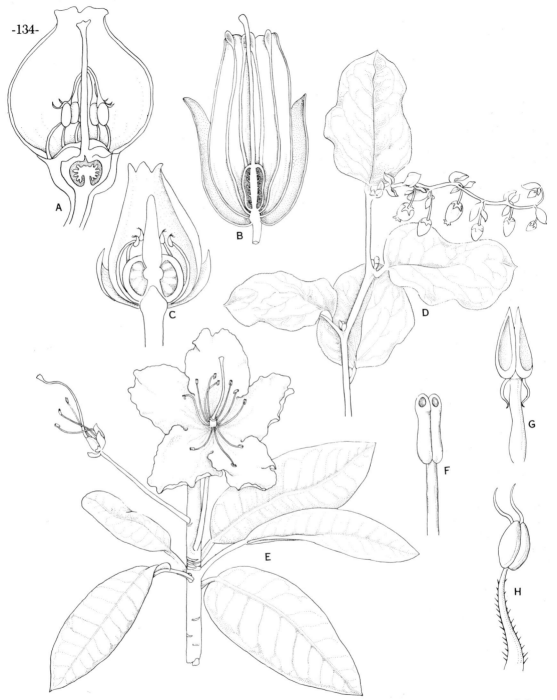

Fig. 53. **Ericaceae.** a, *Vaccinium* flower in longitudinal section; b, *Erica* flower in longitudinal section; c, *Arctostaphylos* flower in longitudinal section; d, *Gaultheria;* e, *Rhododendron;* f, stamen showing anther with terminal pores; g & h, stamens with appendages. (Drawings a-c; f-h after Melchior and Werdermann, 1964).

PYROLACEAE
Pyrola Family

Perennial, somewhat woody herbs, caulescent or acaulescent. Leaves evergreen, foliaceous, alternate or subverticillate, entire or denticulate, exstipulate. Flowers bisexual, actinomorphic, bracteate, in racemes, umbels, cymes, or solitary and scapose. Calyx of 4 or 5 separate or slightly connate sepals. Corolla of 4 or 5 separate petals. Androecium of 10 [8] stamens, opening by terminal pores, pollen united in tetrads. Gynoecium of 4 or 5 united carpels, placentation axile, ovary superior. Fruit a ± globose loculicidal capsule.

4/46; north temperate zone, extending south into Mexico and the West Indies. The three genera listed below are native to the U.S. Of little economic importance; occasionally cultivated as ornamentals.

Selected Genera:

Pyrola (40)-shinleaf

Moneses (1)-*M. uniflora;* flowers solitary

Chimaphila (4)-pipsissewa

Recognition Characters:

Small herbs with evergreen leaves, 4 or 5-merous flowers with separate petals, 10 stamens with anthers opening by terminal pores; fruit a capsule.

$$K^{4-5} C^{4-5} A^{10} \underline{G}^{(4-5)}$$

MONOTROPACEAE
Indian Pipe Family

Parasitic fleshy herbs lacking in chlorophyll. Leaves alternate, reduced to scales, the uppermost often involucrate. Flowers bisexual, actinomorphic, in racemes, heads, or solitary, often dull. Calyx of 2-6 sepals, often bract-like. Corolla of 3-6 free or united petals. Androecium of 6-12 stamens, filaments free or connate, anthers opening by longitudinal slits, pollen grains separate. Gynoecium of 4-6 united carpels, 1-6 locules, stigma capitate. Fruit a loculicidal capsule.

12/30; north temperate areas of both the Old World and New World. Seven genera are native to the U.S. The family is of no economic importance. Mycorrhizal relationships are highly developed in the family.

Selected Genera:

Monotropa (3-4)-Indian pipe, ghost plant

Pterospora (1)-*P. andromedea* is pine drops

Sarcodes (1)-*S. sanguinea* is snow plant

Allotropa (1)-*A. virgata* is sugar stick

Recognition Characters:

Non-green (often red, brown, yellow or white), fleshy, parasitic herbs with scaly leaves; anthers opening by longitudinal slits; loculicidal capsules.

$$K^{2-6} C^{3-6 \text{ or } (3-6)} A^{6-12} \underline{G}^{(4-6)}$$

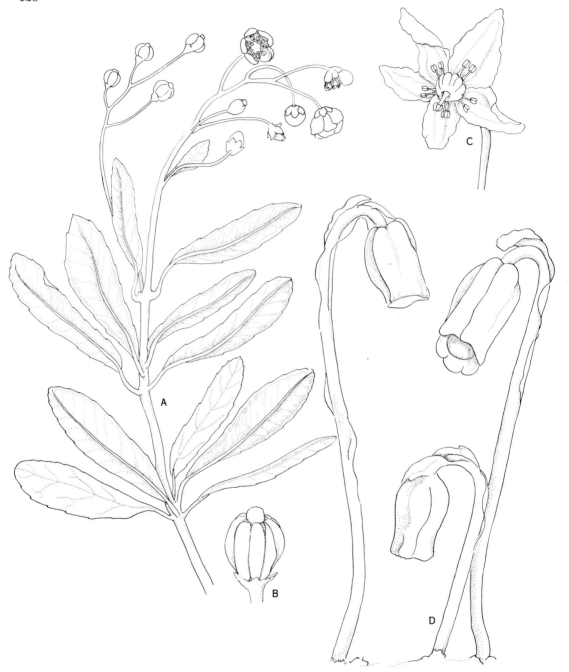

Fig. 54. **Pyrolaceae.** a & b, *Chimaphila* plant and fruit; c, *Pyrola* flower. **Monotropaceae.** d, *Monotropa*.

PRIMULACEAE
Primrose Family

Perennial or annual herbs, rarely sub-shrubs. Leaves generally opposite, whorled, or basal; simple, often gland-dotted, exstipulate. Flowers bisexual, actinomorphic. Calyx of 5 sepals, these persistent. Corolla of 5 petals (united or separate), rarely lacking. Androecium of 5 stamens, these opposite the petals. Gynoecium of 5 united carpels, unilocular, placentation free-central; ovary superior or half-inferior. Fruit a capsule or a pyxis.

28/800; cosmopolitan, but most frequent in the north temperate zones. Eleven genera are native to the U.S., mostly in the eastern part of the country. Of minor economic importance as a source of ornamentals.

Selected Genera:

Primula (540)-primrose; much studied because of heterostyly

Lysimachia (including *Steironema*) (200)-loosestrife

Trientalis (3)-starflower; often 5-7-merous

Cyclamen (20)-cyclamen; popular ornamentals

Dodecatheon (15)-shooting star

Anagallis (40)-pimpernel

Androsace (100)-rock jasmine

Hottonia (2)-an aquatic with dissected leaves

Glaux (1)-*G. maritima* is the sea milkwort of coastal marshes

Recognition Characters:

Herbs with opposite basal, or whorled leaves, 5-merous flowers, stamens opposite the petals, and free-central placentation.

$$K^5 \ C^{(5)} A^5 \ \underline{G}^{(5)}$$

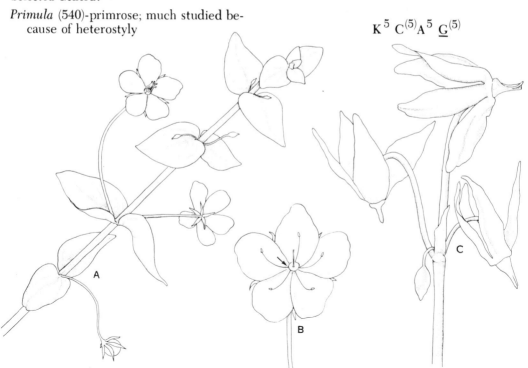

Fig. 55. **Primulaceae.** a & b, *Anagallis* plant and flower detail (arrow indicates stamen insertion opposite a petal); c, *Dodecatheon.*

Magnoliopsid

Order: Family	Common Name	Genera/Species	Distribution	Form	K
Rosales					
Pittosporaceae	Pittosporum	9/200	OW trop.	TSV	5
Hydrangeaceae	Hydrangea	17/250	N. temp. & subtrop.	TS	(4-10)
Grossulariaceae	Gooseberry	1/150	N. temp. & S. Am.	S	(4-5)
Crassulaceae	Stonecrop	35/1500	Cosmo.	HS	4-5
Saxifragaceae	Saxifrage	30/580	Chiefly N. hemis.	H	5
Rosaceae	Rose	100/3000	Cosmo.	TSH	5
Chrysobalanaceae	Chrysobalana	10/400	Trop. & subtrop.	TS	(5)
Leguminosae	Pea	600/13,000	Cosmo.	TSHV	5
Podostemales					
Podostemaceae	River weed	45/130	Chiefly pantrop.	H	2-3?
Haloragales					
Haloragaceae	Water milfoil	6/120	Cosmo.	H	4or4+4
Hippuridaceae	Mare's tail	1/3	Cosmo.	H	0
Myrtales					
Lythraceae	Loosestrife	25/550	Cosmo.	HST	4or6
Trapaceae	Water chestnut	1/30	Euras. & Afr.	H	(4)
Myrtaceae	Myrtle	100/3000	Chiefly Austr.	TS	4-5or(4-!
Punicaceae	Pomegranate	1/2	Subtrop. As.	S	5-8
Onagraceae	Evening primrose	20/650	Temp. & subtrop.	HS	2or4
Melastomataceae	Melastoma	240/3000	Trop. & subtrop.	H[ST]	4-5
Combretaceae	Combretum	20/600	Trop. & subtrop.	TS	5
Proteales					
Elaeagnaceae	Oleaster	3/50	N. hemis.	S	4
Proteaceae	Protea	62/1050	Trop. S. hemis.	TS	(4)

osidae

C	A	G	Fruit Type(s)	Miscellaneous Comments
5	5	(2-5)	cap, ber	Only in cultivation in U.S.
4-5	4-∞	2-5	cap	Often with opposite leaves
5	5	(2)	ber	Often spiny shrubs
4-5	8-10	4-5	fol	Succulent herbs and shrubs
5	5or10	2	cap	Scapose, exstipulate herbs
5	∞	∞,(5),1	ach,drp,pom,fol	Stipulate leaves
0-5z	2-∞	1	drp	Flowers ± zygomorphic
5or5z	10-∞	1	leg	Economically of great importance
0	1-∞	(2-3)	cap	Only on rocks in running water
0	4+4	(2-4)	nut, drp	*Proserpinaca & Myriophyllum* in U.S.
0	1	1	ach	Aquatics with erect stems
4or6	8or12	(2-6)	cap	Hypanthium; petals often crumpled
4	4	(2)	drp	Aquatics; naturalized in U.S.
4-5	∞	(2-3)	cap,ber	Economically very important
5-∞	∞	(∞)	ber	Cultivated in warmer U.S.
2or4	4or8	(4)	cap,ber,nutl	Hypanthium
4-5[z]	8or10	(4-14)	cap, ber	Veins ± parallel; elaborate stamens
5or0	10	1?	drp-like	Four genera native to U.S.
0	4or8	1	ach	Fruit surrounded by fleshy calyx
0	4	1	fol,drp,nut,sam	Widely cultivated ornamentals

Cornales

Rhizophoraceae	Mangrove	16/120	Mostly OW trop.	ST	3-16
Nyssaceae	Nyssa	2/10	N.Am. & E. As.	TS	5or0
Cornaceae	Dogwood	12/100	Widespread	TS[H]	4-5
Garryaceae	Silk tassel	1/18	N. Am.	S	2or4

Santalales

Olacaceae	Olax	25/250	Pantrop.	TSV	4-6
Santalaceae	Sandalwood	30/400	Trop. & temp.	STH	4-5
Viscaceae	Mistletoe	11/450	Cosmopolitan	S	2-4

Rafflesiales

Rafflesiaceae	Rafflesia	8/50	Trop.	H	4-5

Celastrales

Celastraceae	Bittersweet	60/850	Trop. & temp.	TS	3-5
Aquifoliaceae	Holly	2/400	Trop. & temp.	TS	4

Euphorbiales

Buxaceae	Boxwood	6/60	Trop. & subtrop.	STH	4or6
Euphorbiaceae	Spurge	290/7500	Cosmo.	STH	0or5

Rhamnales

Rhamnaceae	Buckthorn	58/900	Cosmo.	TSV	5
Vitaceae	Grape	12/700	Trop. & subtrop.	V	4-5

Sapindales

Staphyleaceae	Bladdernut	5/60	Temp. N. hemis.	TS	5
Sapindaceae	Soapberry	150/2000	Trop. & subtrop.	TS	5
Hippocastanaceae	Horse chestnut	2/15	Mostly N & S Am.	TS	(5)
Aceraceae	Maple	2/200	Mostly N. hemis.	TS	4-5
Burseraceae	Bursera	20/600	Trop.	ST	(3-5)
Anacardiaceae	Cashew	79/600	Chiefly trop.	TSV	5
Simaroubaceae	Quassia	20-32/100-200	Trop. & subtrop.	ST	(3-7)
Rutaceae	Rue	150/1600	Trop. & temp.	ST	5-4
Meliaceae	Mahogany	50/1400	Pantrop.	ST	3-5
Zygophyllaceae	Caltrop	30/250	Trop. & subtrop.	HS[T]	5

3-16	8-∞	$\overline{(2\text{-}4)}$	ber, drp	*Rhizophora mangle* native to S. Fla.
5-8,0	10	$\overline{(1\text{-}2)}$	drp	*Nyssa* native to eastern U.S.
4-5	4-5	$\overline{(2)}$	drp, ber	Only *Cornus* native to U.S.
0	4	$\overline{(2\text{-}3)}$	ber	Unisexual flowers in catkin-like racemes
4-6	4-12	$\underline{(3\text{-}4)}$	drp	*Ximenia & Schoepfia* native to S. Fla.
0	4-5	$\overline{(3\text{-}5)}$	ach, drp	Four genera native to U.S., chiefly in E
0	2-4	$\overline{(3\text{-}4)}$	ber, drp	Tree parasites
0	∞	$\overline{(4\text{-}8)}$	ber	Ours minute stem-parasites in Calif. & Ariz.
3-5	3-5	$\overline{(2\text{-}5)}$	ber,drp,cap,sam	Seed often enveloped in brightly-colored aril
4	4	$\underline{(4)}$	ber, drp	Both genera occur in U.S.
0	4-6	$\underline{(3)}$	cap, drp	*Simmondsia* and *Pachysandra* native to U.S.
0or5	1-∞	$\underline{(3)}$	schizo	Often with a milky latex; many cactus-like
5	5	$\underline{(3)}$	cap, ber	Hypanthium; stamens opposite the petals
$\overline{4\text{-}5}$	4-5	$\underline{(2)}$	ber	Chiefly tendril-bearing vines
5	5	$\underline{(3)}$	cap	Two species of *Staphylea* native to U.S.
5	10	$\underline{(3)}$	cap,nut,ber,drp	Leaves often pinnately compound
5-4z	8-5	$\underline{(3)}$	cap	Ours with opposite, palmately compound leaves
4-5	8	$\underline{(2)}$	schizo or sam	Opposite, palmately lobed leaves
3-5	3-10	$\underline{(2\text{-}5)}$	ber, cap	*Bursera* in the Southeast and Southwest
5	10	$\underline{(3)}$	drp	Leaves pinnately compound, often trifoliolate
3-7	=2C	$\underline{(4\text{-}5)}$	schizo,cap,ber	Six genera native to U.S.
5-4	10or8	$\underline{(5\text{-}4)}$	drp,ber,sam	Glandular punctate leaves
3-5	3-10	$\underline{(2\text{-}5)}$	ber, cap	*Swietenia* native; *Melia* widely naturalized
5	5-15	$\underline{(5)}$	cap,drp-like	Eight genera in U.S.

Geraniales					
Oxalidaceae	Oxalis	8/950	Trop. & subtrop.	H	5
Geraniaceae	Geranium	11/780	Cosmo.	H	5
Limnanthaceae	Meadow foam	2/11	N. Am.	H	3or5
Tropaeolaceae	Nasturtium	2/92	Mex. to S. Am.	H	5z
Balsaminaceae	Touch-me-not	4/500-600	Euras., Afr., N. Am.	H	2+1
Linales					
Linaceae	Flax	12/290	Cosmo.	HS	5
Malpighiales					
Malpighiaceae	Malpighia	60/850	Am. trop.	VST	5
Polygalaceae	Milkwort	13/800	Cosmo.	HSVT	5z
Krameriaceae	Krameria	1/25	Warm Am.	SH	4-5z
Umbellales					
Araliaceae	Ginseng	70/700	Trop.	TS	5
Umbelliferae	Parsley	300/3000	Chiefly N. temp.	H[ST]	5

The Rosidae consist of sixteen orders, 108 families, and about 60,000 species. More than one - third of the Magnoliopsida belong here. Like the Dilleniidae, the subclass is more advanced than the Magnoliidae, but less so than the Asteridae. It is not easily delimited. Most members have flowers with separate petals, one or two ovules per locule, and a well - developed nectary disc.

The orders may be summarized as follows. The Rosales, the most primitive order of the subclass, are defined by a series of tendencies. The Podostemales are small aquatics. The Haloragales are aquatic and terrestrial herbs with a reduced perianth. The Myrtales are best defined by the presence of phloem between the primary xylem and pith. TheCornales are mostly woody plants with epigynous flowers and fleshy fruits. The Santalales show modifications for parasitism. The Celastrales are typically woody plants with polypetalous flowers, nectariferous discs, and one or two ovules per locule. The Euphorbiales have unisexual apetalous flowers, these sometimes aggregated into pseudanthia. The Rhamnales are trees, shrubs, or vines with stamens opposite the petals and well - developed staminal discs. The Sapindales are woody plants, typically with compound leaves. The more advanced families of the Polygalales have zygomorphic flowers with stamens opening by terminal pores. The characters of the Umbellales are those of the Umbelliferae and Araliaceae.

5	10	5	cap, ber	*Oxalis* over much of U.S.
5	10	(5)	schizo	Schizocarp segments hygroscopic
3or5	6or10	3-5	nutl	Endemic to N. Am., mostly in California
5z	8	(3)	schizo	None is native, but *Tropaeolum* widely cultivated
5z	5	(5)	cap	Capsule explosive, hence the common name
5	5or10	(5)	cap	Source of flax and linseed oil
5z	10	(3)	schizo,cap,ber	Five genera native to U.S., most in SW
3z	8	(2)	cap	Superficially papilionaceous flowers
5z	4	(1)	nutl?	Once considered a legume!
5	5	(5)	ber,drp	Three genera native to U.S.
5	5	(2)	schizo	Typically with a compound umbel inflorescence

HYDRANGEACEAE
Hydrangea Family

Shrubs, small trees, sometimes herbs or rarely climbers. Leaves alternate or opposite, simple, dentate, exstipulate. Flowers bisexual, actinomorphic, in cymes, corymbs, or heads. Calyx of 4-10 connate sepals. Corolla of 4-5 [10] free petals. Androecium of 4 to many stamens, in several series. Gynoecium of 2-5 free or partly united carpels, ovary inferior or half-inferior, 3-6 loculed, placentation axile or parietal. Fruit a loculicidal capsule.

17/250; mostly in the northern hemisphere, from the Himalayas to North America. Nine genera are native to the U.S. The family is the source of a few ornamentals. As treated here, the family includes the Philadelphaceae and Iteaceae.

Selected Genera:

Hydrangea (23)- hydrangea; widely cultivated

Philadelphus (71)- mock orange; commonly cultivated

Itea (20)- sweet spire, Virginia willow

Carpenteria (1)- *C. californica* is the tree anemone

Deutzia (60)- deutzia, widely planted ornamentals

Whipplea (1)- *W. modesta* is the yerba de selva of the Pacific coast

Recognition Characters:

Shrubs with opposite exstipulate leaves, flowers with numerous stamens and a wholly or partially inferior ovary.

$$K^{(4-10)} \; C^{4-5} \; A^{4-\infty} \; \overline{G}^{2-5}$$

Fig. 56. **Hydrangeaceae.** a, *Philadelphus;* b, *Hydrangea* (arrow indicates fertile flowers).

GROSSULARIACEAE
Gooseberry Family

Shrubs, often with spines. Leaves alternate, simple and variously lobed, often fascicled, with or without stipules. Flowers bisexual or unisexual (the species dioecious), actinomorphic, in racemes or subsolitary. Calyx of 4-5 connate, persistent sepals, sometimes petaloid. Corolla of [4] 5 small, scale-like petals. Androecium of 5 stamens, rarely 4. The lower portions of the perianth and androecium fused to form a rotate to tubular hypanthium. Gynoecium of 2 united carpels, unilocular, placentation parietal, the ovary inferior. Fruit a berry

1/150; north temperate zone and South America. *Ribes* is native to the U.S. A source of ornamentals and edible fruits.

Genus:

Ribes (including *Grossularia*) (150)- the gooseberries (Subgenus *Grossularia*) have spines at the nodes, bristles on the internodes, and the fruits do not disarticulate from the pedicels; currants (Subgenus *Ribesia*) are without nodal and internodal spines and bristles, and the fruits disarticulate from the pedicels

Recognition Characters:

Spiny shrubs with alternate lobed leaves, 5-merous flowers, a petaloid persistent calyx, rotate to tubular hypanthium, inferior ovary, and spiny or spinless berry.

$$\underline{K}^{(4-5)}\ C^5\ A^5\ \overline{G}^{(2)}$$

Fig. 57. **Grossulariaceae.** Habit, fruit, and flower of *Ribes*.

CRASSULACEAE
Stonecrop or Orpine Family

Succulent herbs and subshrubs. Leaves alternate or opposite, simple, fleshy, exstipulate. Vegetative reproduction common from rhizomes, offsets, and bulbils. Flowers usually perfect. Calyx of 4-5 [30] free sepals. Corolla of as many petals as sepals, separate or connate. Androecium of as many or twice as many stamens as petals. Gynoecium of 3 to several free or basally connate carpels, each subtended by a glandular scale, ovary superior, unilocular, placentation parietal. Fruit a follicle.

35/1500; cosmopolitan, but centered in South Africa; essentially absent from Australia and South America. Nine genera are native to the U.S. Of no real economic importance except as a source of ornamentals, many of them quite curious. Generic delimitations within the family often seem arbitrary.

Selected Genera:

Crassula (including *Tillaea*) (300)- stonecrop; many cultivars

Sedum (500)- stonecrop, orpine; many in cultivation

Sempervivum (30)- houseleek, live-forever

Aeonium (40)- widely cultivated; even leaves are ornamental

Echeveria (100)- a commonly cultivated succulent

Dudleya (84)- live-forever

Cotyledon (44)- commonly cultivated

Kalanchoë (including *Bryophyllum*) (200)- air plant; plantlets are asexually produced along the leaf margins

Recognition Characters:

Succulent herbs and subshrubs with 4-5-merous flowers, ± separate carpels, each subtended by a scale-like gland.

$$K\ 4\text{-}5\ \ C\ 4\text{-}5\ \ A\ 8\text{-}10\ \ \underline{G}\ 4\text{-}5$$

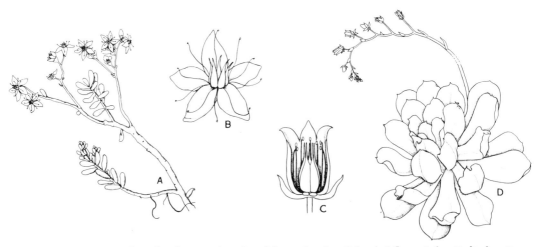

Fig. 58. **Crassulaceae.** a & b, *Sedum* flowering branch and flower detail; c, *Kalanchoë* flower in longitudinal section showing tubular corolla; d, *Dudleya*.

SAXIFRAGACEAE
Saxifrage Family

Perennial, rarely annual scapose herbs. The leaves alternate, basal, exstipulate. Flowers bisexual, actinomorphic, in racemose or cymose inflorescences, rarely solitary. Calyx often of 5 sepals. Corolla of 5 petals (rarely 0), often clawed. Androecium of 5 or 10 stamens. Gynoecium often of 2 carpels, united below, or more rarely 5 free carpels; ovary usually superior, the flowers perigynous, placentation axile. Fruit a capsule.

30/580; chiefly of the cooler and temperate regions of the northern hemisphere; a few are southern hemisphere. About twenty genera are native to the U.S., more species occurring in the West than any other part of the country. Several are cultivated as ornamentals. The family closely resembles the Rosaceae (see Table 8 below). As treated here the Saxifragaceae includes the Penthoraceae and Parnassiaceae, but excludes the subfamilies Grossularioideae and Hydrangeoideae of earlier workers.

Selected Genera:

Saxifraga (350)- saxifrage; many in cultivation

Bergenia (8)- bergenia, a popular ornamental

Mitella (15)- miterwort, bishop's cap

Penthorum (4)- ditch stonecrop; often placed in Crassulaceae

Tiarella (6)- foam flower

Heuchera (50)- alum root

Chrysosplenium (55)- golden saxifrage; apetalous

Astilbe (30-35)- widely cultivated

Recognition Characters:

Perennial scapose herbs with alternate, basal, exstipulate leaves; 5-merous flowers with 2-5 imperfectly united carpels, the ovary borne within a hypanthium, perigynous. The Saxifragaceae, Rosaceae, and Crassulaceae form a complex of families that are not easily distinguished.

$$\underline{K}^5 \underline{C}^5 \underline{A}^{5 \text{ or } 10} \underline{G}^2$$

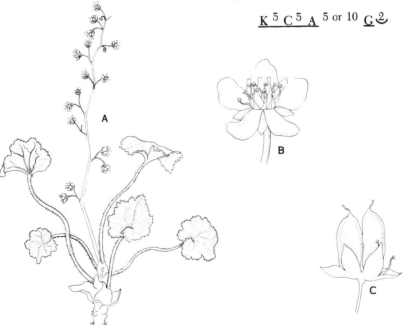

Fig. 59. **Saxifragaceae.** a-c, *Saxifraga* plant, flower, and fruit.

Table 9. Comparison of the Saxifragaceae, Crassulaceae, and Rosaceae.

Family	Stipules	Ovary Position	Carpels	Stamens	Endosperm	Placentation	Fruit
Saxifragaceae	–	perigynous	2	5-10	abundant	axillary	capsule
Crassulaceae	–	hypogynous	5	8-10	scant	parietal	follicle
Rosaceae	+	peri-, epi-	1,5, ∞	∞	rare	axile, marginal	achene pome drupe

ROSACEAE
Rose Family

Trees, shrubs, or herbs. Leaves alternate, simple or compound, usually stipulate. Flowers bisexual, rarely unisexual (the species then dioecious), actinomorphic, perigynous. Calyx usually of 5 connate sepals. Corolla of 5 [0] petals, arising from the rim of the hypanthium. Androecium usually of numerous stamens in 1- several series, sometimes as few as 5 or 10 stamens. Gynoecium unicarpellate, apocarpous, or syncarpous; ovary superior (situated within a free hypanthium) or inferior (ovary borne within an adnate hypanthium). Fruit an achene, aggregate, drupe, pome, or follicle, depending upon the carpel number and ovary position.

100/3000; cosmopolitan, but particularly common in Europe, Asia, and North America. About fifty genera occur in the U.S. The family is of considerable economic importance because of edible fruits (many derived from *Prunus*) and many ornamentals. As treated here, the family includes the Spiraeaceae, Pomaceae, Malaceae, Poteriaceae, Amygdalaceae, and Drupaceae.

Selected Genera:

Subfamily: Rosoideae (stipulate trees, shrubs, and herbs; gynoecium apocarpous, ovary superior, the flower perigynous; carpels several to many, rarely a few to 1; fruit an achene, drupelet or aggregate of these)

Rosa (100-200)- rose

Filipendula (10)- meadowsweet

Rubus (700+)- bramble, blackberry, raspberry, dewberry

Sanguisorba (30)- burnet

Adenostoma (2)- chamise

Cercocarpus (10)- mountain mahogany

Purshia (2)- antelope bush

Fallugia (1)- apache plume

Geum (56)- avens

Potentilla (300+)- cinquefoil

Fragaria (20-30)- strawberry

Subfamily: Spiraeoideae (exstipulate shrubs; gynoecium apocarpous, ovary superior, the flower perigynous; carpels 2-5, rarely 1 to 12; fruit a series of follicles, a capsule or rarely an achene)

Spiraea (100)- spirea; widely cultivated shrubs

Gillenia (2)- Indian physic, American ipecac

Aruncus (3)- goat's beard

Holodiscus (8)- ocean spray

Physocarpus (10)- ninebark

Subfamily: Prunoideae (= Amygdaloideae) (stipulate shrubs and trees; gynoecium unicarpellate, rarely 2-5 carpels, the ovary superior, the flower perigynous;

fruit usually a solitary drupe

Prunus (200)- stone fruits, plum, apricot, peach, almond, cherry

Oemleria (=*Osmaronia*) (1)-*Oe. cerasiformis* is the oso berry

Subfamily: Pomoideae (= Maloideae = Pyroideae) (stipulate shrubs and trees; gynoecium syncarpous, ovary inferior, the flower epigynous; carpels 2-5; fruit a pome

Pyrus (25)- pear

Malus (25)- apple

Cydonia (1)- *C. oblonga* is the quince

Sorbus (100+)- mountain ash; ornamental

Eriobotrya (25-30)- loquat; edible fruits

Amelanchier (20-25)- service berry, shad bush

Cotoneaster (95)- cotoneaster; ornamentals

Pyracantha (8-10)- firethorn; ornamental

Crataegus (200)- hawthorne

Recognition Characters:

Stipulate herbs, shrubs, and trees with 5-merous flowers, a hypanthium, and many stamens. The four subfamilies are compared below.

$$\underline{K}\ 5\ C\ 5\ A\ ^\infty\ \underline{G}\ 1\ \text{or}\ \underline{G}\ ^\infty\ \text{or}\ \overline{\underline{G}}\ (5)$$

Table 10. Comparison of the subfamilies of the Rosaceae.

Subfamily	Stipules	Carpel No.	Ovary Position	Fruit
Rosoideae	Present	∞	Superior-Perigynous	drp, ach
Spiraeoideae	Absent	2-5 [1-12]	Superior-Perigynous	fol, cap
Prunoideae	Present	1	Superior-Perigynous	drupe
Pomoideae	Present	2-5	Inferior-Epigynous	pome

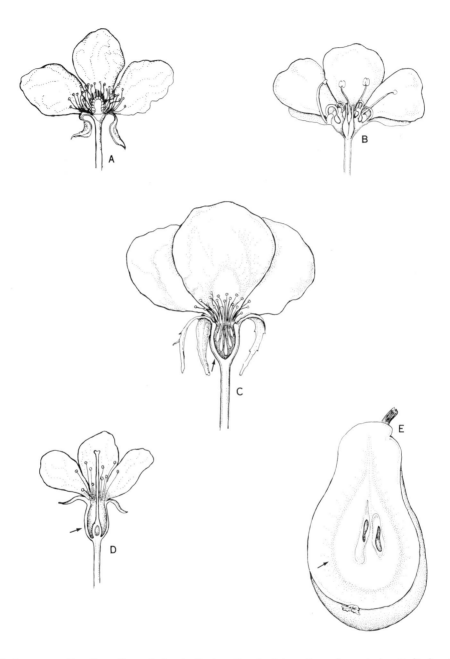

Fig. 60. **Rosaceae**. a, Rosoideae, flower in longitudinal section; b, Spiraeoideae, flower in longitudinal section; c, Rosoideae, flower in longitudinal section (arrow indicates leathery hypanthium); d, Prunoideae, flower in longitudinal section (arrow indicates hypanthium); e, Pomoideae, fruit in longitudinal section (arrow indicates region of transition from ovary tissue to hypanthium).

Fig. 61. **Rosaceae.** a, *Fragaria;* b, *Pyracantha;* c, *Rubus.*

LEGUMINOSAE (= FABACEAE)
Pea Family

Trees, shrubs, herbs, and vines. Roots often with bacterial nodules. Leaves alternate, once pinnately compound, twice pinnately compound, palmately compound, or simple through reduction, mostly stipulate. Flowers bisexual, actinomorphic or zygomorphic, in racemes, heads, umbels, or panicles. Calyx of 5 connate sepals. Corolla of 5 petals, [1,0], distinct or the lower two petals connate. Androecium usually of 10 stamens, sometimes fewer, sometimes numerous; free, monadelphous, or diadelphous. Gynoecium unicarpellate, unilocular, placentation marginal. Fruit typically a legume or loment, but its gross appearance quite variable (straight, curved, or coiled; dry or fleshy; smooth or spiny; inflated or not; dehiscent along one or two sutures or indehiscent).

The Leguminosae are an exceedingly large family of 600 genera and 13,000 species, perhaps 3rd ranked in size. The family is traditionally treated as one large, somewhat heterogeneous taxon. A few recent authors have recognized the three subfamilies as distinct families. As treated here, the family includes the Mimosaceae, Caesalpiniaceae, and Papilionaceae.

The 3 subfamilies may be distinguished by the following characters. Further details of floral structure, size, distribution, and economic products are presented under the individual treatments of the subfamilies.

Subfamily: Mimosoideae

Trees and shrubs, rarely herbs, often with bipinnately compound leaves. Flowers actinomorphic, 5-merous, petals distinct; stamens (4)10-many; gynoecium unicarpellate; commonly in tight heads with numerous, small flowers, each with numerous radiating stamens with long, often brightly colored filaments.

40/2000; mostly tropical and subtropical.

Economically important as a source of numerous ornamentals, several valuable timber trees, and gum arabic.

Selected Genera:

A. Stamens more than 10
  Tribe: Ingleae (stamens basally fused)

  Albizia (100-150)- silk tree, mimosa tree
  Calliandra (150)- fairy duster; widely cultivated
  Pithecellobium (150-200)- guaymochil

  Tribe: Acacieae (stamens free)
  Acacia (700-800)- acacia, wattle; many in cultivation

A. Stamens as many as or twice as many as the number of petals
  Tribe: Mimoseae (anthers glandless in bud)
  Schrankia (30)- sensitive brier, a vine
  Mimosa (450-500)- mimosa, sensitive plant
  Desmanthus (40)- prairie mimosa

  Tribe: Adenanthereae (anthers with glands in bud)
  Prosopis (35)- mesquite, screw bean

Subfamily: Caesalpinioideae

Trees or shrubs, rarely herbs. Leaves pinnate, bipinnate, or simple. Flowers often large and showy. Corolla ± zygomorphic, although not as markedly as the Papilionoideae; the adaxial (upper) petal inside the 2 lateral ones; stamens usually 10, free.

150/2200; primarily of the old world tropics. Several are prized ornamentals, a few have edible fruits, several are important tropical timber trees, and *Haematoxylon* yields the valuable biological dye.

Selected Genera:

A. Leaves simple or 1-pinnately compound

  Tribe: Cercideae (= Bauhinieae) (leaves simple)

  Cercis (7)-redbud, Judas tree

  Bauhinia (250)-orchid tree; leaves bilobed

  Tribe: Cassieae (leaves 1-pinnate)

  Cassia (500)- senna

  Ceratonia (1)- *C. siliqua* is St. John's bread

A. Leaves typically twice-pinnately compound

  Tribe: Caesalpinieae

  Gleditsia (15)- honey locust; armed with stout thorns

  Gymnocladus (2)- *G. dioica* is the Kentucky coffee bean tree

  Parkinsonia (2)- palo verde of the Southwest

  Cercidium (10)- palo verde of the Southwest

  Delonix (4)-*D. regia* is the royal poinciana tree

  Caesalpinia (125)- poinciana; widely cultivated

Subfamily: Papilionoideae (= Lotoideae = Faboideae)

Herbs, shrubs, or trees. Leaves pinnately compound, palmately compound, or simple. Flowers typically markedly zygomorphic. Calyx of 5 united sepals. Corolla of 5 unequal petals, the adaxial (upper) petal outside the 2 lateral ones. The uppermost, outer petal is called the standard or banner. It often appears to be 2 petals because of lobing and pigmentation. The 2 lateral petals are clawed and similar to one another. These are referred to as wing petals. The two basal petals, the keel petals, are often united into a boat - shaped structure, the keel. The androecium and gynoecium lie within the keel. A corolla of this configuration is termed papilionaceous. Caesalpinoid legumes, although zygomorphic, do not typically reach this level of elaboration. One exception is the almost papilionaceous corolla of *Cercis*.

400/9000; found primarily in the temperate regions of both the northern and southern hemisphere. The subfamily is of tremendous economic importance, perhaps third after the Gramineae and Palmae.

Selected Genera:

A. Stamens free or nearly so

  Tribe: Sophoreae (leaves pinnate or simple; leaves with a joint between the petiole and lamina)

  Sophora (55-60)- Japanese pagoda tree; mescal bean

  Cladrastis (4)- yellow wood, an ornamental tree

  Tribe: Podalyrieae (leaves palmately compound or simple; leaves not jointed)

  Thermopsis (20)- false lupine

  Baptisia (30)- false indigo, wild indigo

A. Stamens monadelphous or diadelphous

  B. Legumes dehiscing transversely

  Tribe: Hedysareae

  Coronilla (25)- crown vetch

  Arachis (12)- *A. hypogaea* is the peanut or goober

  Desmodium (200)- tick trefoil

  Lespedeza (100)- bush clover, lespedeza

  B. Legumes dehiscing longitudinally or indehiscent

  Tribe: Genisteae (shrubs; leaves simple or more commonly palmately trifoliolate)

Crotalaria (500)- rattle box

Lupinus (200)- lupine, bluebonnet; many are toxic

Spartium (1)- *S. junceum* is the Spanish broom

Genista (90)- broom

Ulex (15)- *U. europaeus* is gorse, an aggressive weed

Cytisus (60)- Scotch broom, French broom

Tribe: Galegeae (leaves usually pinnately compound; rachis not ending in tendrils)

Indigofera (500)- indigo; the source of the dye

Psoralea (150)- plants gland-dotted

Amorpha (20)- lead plant

Dalea (250)- indigo bush

Petalostemon (40)- prairie clover

Wisteria (9)- wisteria, an ornamental vine

Robinia (20)- black locust

Astragalus (1,600)- locoweed; toxic

Oxytropis (200-300)- locoweed; several are toxic

Glycyrrhiza (15)- locorice

Tribe: Loteae (leaves pinnately 3-many foliolate)

Lotus (150)- bird's foot trefoil

Tribe: Trifolieae (usually herbs; leaves pinnately or palmately trifoliolate)

Medicago (110)- alfalfa, medic

Melilotus (25)- sweet clover

Trifolium (300)- clover

Tribe: Fabeae (= Vicieae) (leaves pinnately compound; rachis ending in a tendril)

Cicer (15-20)- chick pea, garbanzo

Vicia (150-200)- vetch

Lathyrus (130)- sweet pea, vetchling

Lens (6)- *L. culinaris* is the lentil

Pisum (6-7)- pea

Tribe: Abreae (woody plants; stamens 9)

Abrus (4)- *A. precatorius* is the precatory or pater noster bean, a toxic ornamental

Tribe: Phaseoleae (often twining plants; leaves pinnately trifoliolate)

Glycine (60)- *G. max* is the soybean

Erythrina (104)- coral tree

Apios (8)- ground nut

Pueraria (15)- kudzu vine; an aggressive vine!

Phaseolus (200)- beans; many varieties in cultivation

Vigna (150)- cow pea, blackeyed pea

Recognition Characters:

Herbs, shrubs, vines, and trees with alternate, stipulate, compound leaves. Flowers actinomorphic or zygomorphic, 5-merous, unicarpellate; fruit a legume or loment. See Table 11.

Mimo. $K^{(5)} C^5 A^{[4]10-\infty} \underline{G}^1$

Caesal. $K^{(5)} Cz^5 A^{10} \underline{G}^1$

Papilio. $K^{(5)} Cz^{1+2+\underline{2}} A^{\underline{9}+1[10]} \underline{G}^1$

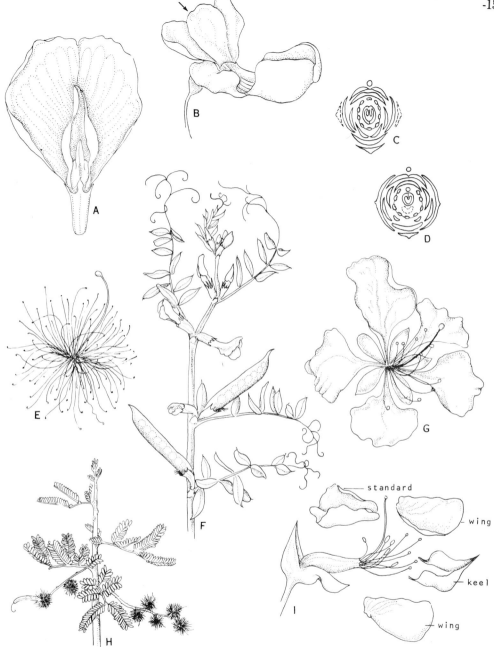

Fig. 62. **Leguminosae.** a, papilionoid flower; b, caesalpinioid flower (arrow indicates standard positioned between wing petals); c, generalized papilionoid flower in x-s (note that the standard is outside the wing petals); d, generalized caesalpinioid flower in x-s (note that the banner is inside the wing petals); e, mimosoid flower; f, *Lathyrus* flowers, fruits, and tendrils; g, *Delonix*, a caesalpinioid; h, *Acacia* leaves, flowers, and fruit; i, exploded dissection of a papilionoid flower.

Table 11. Comparison of the subfamilies of the Leguminosae.

character	Mimosoideae	Caesalpinioideae	Papilionoideae
symmetry	actinomorphic	± zygomorphic	zygomorphic
petals	valvate	banner inside wings	banner outside wings
stamen no.	[4]10-many	10 [many]	10, often 9+1
pollen	compound	simple	simple
leaves	bipinnate	pinnate [simple]	pinnate, palmate [simple].

LYTHRACEAE
Loosestrife Family

Herbs, shrubs, or trees. Leaves opposite or whorled, simple, entire, stipules small or lacking. Flowers bisexual, actinomorphic or zygomorphic, perigynous. Calyx of 4 or 6 [8] sepals. Corolla of as many petals as sepals or absent. Androecium usually of twice as many stamens as petals, in 2 series of different lengths. Gynoecium of 2-6 united carpels, as many locules, ovary superior, situated within a free hypanthium, placentation axile. Fruit a capsule, opening by valves, a transverse slit, or irregularly.

25/550; widely distributed. Seven genera are native to the U.S. A source of dyes and ornamentals.

Selected Genera:

Lythrum (30)- loosestrife

Lagerstroemia (30)- crepe myrtle; an ornamental tree

Cuphea (200)- cigarflower

Lawsonia (1)- *L. inermis* is the source of henna, a dye

Recognition Characters:

Herbs and woody plants with 4- or 6-merous flowers with a hypanthium, often crumpled corolla and stamens twice the number of petals; stamens in two sets, one long, the other short.

$$\underline{K} \; ^{4 \text{ or } 6} \; C \; ^{4 \text{ or } 6} \; \underline{A} \; ^{8 \text{ or } 12} \; \underline{G} \; (2\text{-}6)$$

MYRTACEAE
Myrtle Family

Trees or shrubs. Leaves opposite, simple, entire, glandular punctate. Flowers bisexual, actinomorphic, epigynous. Calyx of 4 or 5 sepals, free or connate, sometimes forming a lid which drops from the flower. Corolla of 4-5 petals. Androecium of numerous stamens. Gynoecium of 2-3 united carpels, ovary often inferior. Fruit a loculicidal capsule or berry, rarely a drupe.

100/3000; chiefly in Australia and tropical America. Three genera are native to the U.S., mostly in Florida. Although not widely known in this country, the family is economically important as a source of handsome ornamentals, food plants, wood, and condiments.

Selected Genera:

Subfamily: Leptospermoideae (leaves opposite or alternate; fruit a capsule)

Eucalyptus (605)- eucalyptus; timber and ornamentals

Leptospermum (25)- Australian tea tree

Callistemon (12)- bottlebrush, an ornamental

Melaleuca (101)- cajeput tree, punk tree

Subfamily: Myrtoideae (leaves strictly opposite; fruit a drupe or berry)

Calyptranthes (101)- spikewood; Florida

Eugenia (600)- cloves

Psidium (100)- guava

Pimenta (14)- allspice, oil of bay rum

Recognition Characters:

Woody plants with glandular punctate leaves, numerous stamens, and an inferior ovary.

$$K \; ^{4\text{-}5} \; C \; ^{4\text{-}5} \; A \; ^{\propto} \; \overline{G} \, (2\text{-}3)$$

ONAGRACEAE
Evening Primrose Family

Herbs, rarely shrubs or trees. Leaves alternate or opposite, simple, exstipulate; if stipulate, these quickly falling. Flowers bisexual, actinomorphic, rarely zygomorphic; in spikes, racemes, panicles, or solitary; epigynous with a hypanthium; 4-merous. Calyx of 4 [2, 3, or 5] sepals, distinct and inserted on the rim of the hypanthium. Corolla of 4 [2 or more, rarely 0] clawed petals. Androecium of as many stamens as petals or twice as many. Gynoecium of 4 [2 or 5] united carpels, ovary inferior, placentation axile. Fruit usually a loculicidal capsule, sometimes a berry or nutlet.

20/650; primarily temperate and subtropical. About a dozen genera are native to the U.S. Of little economic importance; a few are ornamentals. As treated here, the family includes the Oenotheraceae and Epilobiaceae.

Selected Genera:

Ludwigia (30)- water primrose, seedbox

Zauschneria (3)- California fuchsia

Epilobium (200)- fireweed, willow herb

Oenothera (80)- evening primrose

Camissonia (55)- evening primrose (included in *Oenothera* by many early workers)

Clarkia (ca. 30)- several are natives; a few in cultivation

Fuchsia (100)- fuchsia; a very popular ornamental shrub

Circaea (7)- flowers 2-merous; enchanter's nightshade

Lopezia (16)- an ornamental with zygomorphic flowers

Recognition Characters:

4-merous herbs (more infrequently 2-merous) with an interior ovary and hypanthium.

$$\underline{K}^{\,4}\ \underline{C}^{\,4}\ \underline{A}^{\,8}\ \overline{G}^{\,(4)}$$

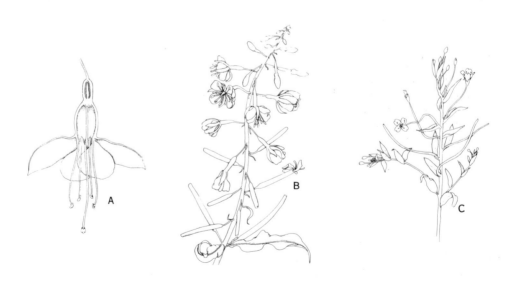

Fig. 63. **Onagraceae.** a, *Fuchsia* flower in longitudinal section; b & c, *Epilobium* flowers and fruits.

CORNACEAE
Dogwood Family

Trees, shrubs, rarely perennial herbs. Leaves opposite or alternate, simple, entire, exstipulate. Flowers bisexual or unisexual (the species monoecious, dioecious, or polygamous), actinomorphic, in cymes or panicles, small, sometimes in heads subtended by an involucre, thus appearing as a single flower. Calyx of 4 or 5 [0] small sepals. Corolla of 4, 5, or 0 distinct petals. Androecium of 4 or 5 stamens. Gynoecium of [1] 2 [4] united carpels, ovary inferior, placentation axile. Fruit usually a drupe, sometimes a berry.

12/100; widespread, but scattered distribution, primarily throughout the temperate regions of the northern and southern hemispheres. A single genus, *Cornus*, occurs in the U.S. Of little economic importance; many are garden ornamentals.

Selected Genus:

Cornus (45)- dogwood; handsome native trees and ornamentals.

Recognition Characters:

Trees and shrubs, often with opposite leaves, arcuate veins, 4-5-merous flowers, inferior ovary, and a drupe or berry fruit type.

$$K^{4-5} C^{4-5} A^{4-5} \overline{G}^{(2)}$$

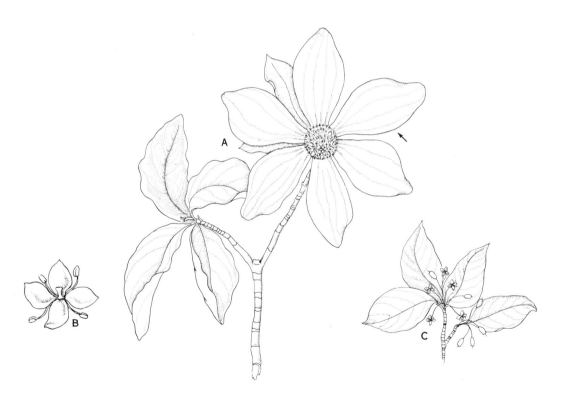

Fig. 64. **Cornaceae.** *Cornus.* a, arrow indicates one of six petaloid bracts subtending the inflorescence; b, flower detail; c, a species without conspicuous floral bracts.

GARRYACEAE
Silk Tassel Family

Shrubs with 4-angled stems. Leaves opposite, entire, exstipulate, evergreen, petioles connate at the base. Flowers unisexual (the species dioecious), in silky catkin-like racemes, 1-3 flowers in the axil of subtending bracts. Staminate flowers with 4 sepals, no petals, and 4 stamens. Pistillate flowers with no perianth, or with 2-4 minute calyx teeth at the summit of the ovary, and 2-3 united carpels, ovary inferior. Fruit a berry.

1/18; North America and the West Indies. Sometimes cultivated as an ornamental.

Genus:

Garrya (18)- silk tassel tree; western and southwestern areas

Recognition Characters:

Dioecious shrubs with exstipulate opposite leaves and catkin-like pendulous inflorescences.

$$K\ 4\ C\ 0\ A\ 4\ G\ 0 \quad \text{(staminate flowers)}$$

$$K\ 0\ \text{or}\ 2\text{-}4\ C\ 0\ A\ 0\ \overline{G}\ (2\text{-}3) \quad \text{(pistillate flowers)}$$

VISCACEAE
Mistletoe Family

Parasitic shrubs of tree branches. Leaves typically opposite, leathery, the nerves ± parallel. Flowers small, bisexual or unisexual. Perianth of 2-4 tepals, inserted on a cup-shaped receptacle, the calyx and corolla not differentiated. Androecium of as many stamens as perianth segments. Gynoecium unilocular, variously interpreted as unicarpellate or 3-4 carpellate; the ovary inferior; ovules seemingly absent. Fruit a drupe or berry, with a viscous layer within the vascular bundles.

11/450; cosmopolitan. *Arceuthobium* and *Phoradendron* are native to the U.S. The mistletoe of Europe and America has seasonal economic importance. The Viscaceae are traditionally treated as a subfamily of the Loranthaceae.

Selected Genera:

Viscum (65)- *V. album* is the European mistletoe

Phoradendron (300)- *P. flavescens* is the U.S. mistletoe

Arceuthobium (15)- dwarf mistletoe

Recognition Characters:

Shrubby parasites of the branches of oaks, alders, cottonwoods, and conifers.

CELASTRACEAE
Bittersweet Family

Trees, shrubs, often climbing. Leaves opposite or alternate, simple. Flowers bisexual, actinomorphic, small, in cymes or fascicles. Calyx of 3-5 sepals. Corolla of 3-5 [0] petals. Stamens 3-5 [10]. Inserted on or below the margin of a disc. Gynoecium of 2-5 united carpels partially sunken in the disc, the ovary superior. Fruit a berry, drupe, samara, loculicidal capsule, or indehiscent capsule. Seeds often with a brightly colored aril.

60/850; widespread in tropical and temperate regions. Ten genera are native to the U.S. Several are cultivated as ornamentals.

Selected Genera:

Celastrus (35)- bittersweet

Euonymus (220)- burning bush, wahoo, spindle tree

Paxistima (5)- Oregon boxwood

Recognition Characters:

Shrubs, often climbing. Stamens arising from a disc which may be adnate to the gynoecium. Seeds often enveloped by a brightly colored aril.

$$K\ 3\text{-}5\ C\ 3\text{-}5\ A\ 3\text{-}5\ \overline{G}\ (2\text{-}5)$$

AQUIFOLIACEAE
Holly Family

Trees or shrubs, often evergreen. Leaves alternate, simple, often spiny. Flowers bisexual or unisexual (the species dioecious or polygamous), actinomorphic, small, greenish. Flowers with 4 sepals, 4 petals, 4 stamens, and 4 united carpels, ovary superior. Fruit a drupe.

2/400; widespread, but rare in Africa and Australia. The two genera listed below are native to the U.S. *Ilex* is of some importance as a source of wood and many ornamentals.

Genera:

Ilex (including *Bryonia*) (400)- holly, yaupon, yerba maté

Nemopanthus (1)- *N. mucronata* is the mountain holly, a northeastern U.S. endemic

Recognition Characters:

Woody plants, often with spiny or toothed leaf margins. Flowers axillary, 4-merous.

$$K^4 C^4 A^4 \underline{G}^{(4)}$$

EUPHORBIACEAE
Spurge Family

Shrubs, trees, and herbs, rarely lianas. Many are xerophytic and cactoid; most with a milky latex. Leaves alternate, sometimes opposite or whorled, simple or compound, usually stipulate, but stipules may be modified into glands, hairs, or spines. Flowers unisexual (the species usually monoecious); actinomorphic; calyx and corolla present, or either or both lacking; variously disposed in racemose or cymose inflorescences. Fruit a schizocarp.

Two major floral patterns are evident within the family; one exhibited by *Euphorbia* and its relatives (here referred to as the **euphorbia type**) and that of the remaining genera (**non-euphorbia type**).

Euphorbia type

Flowers unisexual, borne in a complex, highly reduced cymose inflorescence, the **cyathium**. The cup-like portion of the cyathium, the **involucre**, usually contains several staminate flowers and a single pistillate flower. The staminate flower is reduced to a single stamen. The pistillate flower is reduced to a tricarpellate, trilocular gynoecium with a superior ovary. Both kinds of flowers are pedicellate. Each locule contains a single carunculate seed which is often mottled. The styles are often bifid.

$$K^0 C^0 A^1 G^0 \quad \text{(staminate flower)}$$

$$K^0 C^0 A^0 \underline{G}^{(3)} \quad \text{(pistillate flower)}$$

Non-euphorbia type

Flowers unisexual, the species monoecious or dioecious, usually actinomorphic. Perianth 5-merous, with one or both of the perianth whorls missing. Staminate flowers with or without perianth, with 5 or 10 [sometimes numerous, to 1000] free or variously united stamens. Pistillate flowers with or without perianth, sometimes with staminodia, and a trilocular [2 or 4] carpellate gynoecium with a superior ovary. One or rarely 2 carunculate, often mottled seeds per locule; styles often bifid.

$$K^{0 \text{ or } 5} C^0 A^{1-\infty} G^0 \quad \text{(staminate flower)}$$

$$K^{0 \text{ or } 5} C^{0 \text{ or } 5} A^0 \underline{G}^{(3)} \quad \text{(pistillate flower)}$$

290/7500; particularly well represented in tropical Africa and tropical America. About 25 genera occur in the U.S., particularly in the Southeast. Valuable economic products include: Para rubber, ceara rubber, castor oil, tung oil, tapioca, and cassava. Many are cultivated as handsome ornamentals in warmer parts of the world. Most members of the family are poisonous and should be treated with great respect. Avoid contact with the milky latex. The eyes and lining of the mouth and throat can become badly inflamed and swollen through contact.

Selected Genera:

Subfamily: Phyllanthoideae (ovules 2/locule; no latex)

Phyllanthus (480)- otaheite gooseberry, emblic

Breynia (30)- snow bush; an ornamental

Subfamily: Euphorbioideae (= Crotonoideae) (1 ovule/locule; latex present)

Euphorbia (1,600)- spurge; many weeds and ornamentals

Croton (700)- several native species

Aleurites (15)- tung oil tree; of declining economic importance

Hevea (20)- *H. brasiliensis* is the Pará rubber tree

Acalypha (430)- chenile plant, redhot-cattail; ornamental

Ricinus (1)- *R. communis* is the castor bean; seeds lethal

Codiaeum (14)- croton, handsome ornamentals

Jatropha (150)- coral plant, an ornamental

Manihot (1)- *M. esculenta* is the cassava, manioc or tapioca plant

Cnidoscolus (75)- tread softly, mala mujer; stinging hairs!

Sapium (100)- Chinese tallow tree

Hippomane (5)- *H. mancinella* is the manchineel tree

Pedilanthus (14)- red bird cactus

Recognition Characters:

Herbs and shrubs with milky latex, unisexual often highly reduced flowers with 3-carpellate gynoecium and mottled carunculate seeds.

Fig. 65. **Euphorbiaceae.** a, *Euphorbia* (arrow indicates cyathium); b, generalized cyathium; c, *Euphorbia*, a cactoid species; d, composite cyathium showing petaloid appendage, horn-like appendage, and glandular appendage on the rim of the involucre. (Drawing b after Porter, 1967).

RHAMNACEAE
Buckthorn Family

Trees, shrubs, and vines; rarely herbs. Leaves alternate or opposite, simple, stipulate. Flowers actinomorphic, perigynous, bisexual, small, greenish or brightly colored. Calyx of 5 [4] separate sepals. Corolla of 5 [4,0] small separate petals. Androecium of 5 [4] stamens, these opposite the petals and inserted below a staminal disc. Gynoecium of 3 or 2 united carpels, as many locules, the placentation basal. Fruit a capsule or drupe-like berry.

58/900; cosmopolitan. Ten genera are native to the U.S. Economically of little importance; edible fruits are derived from the jujube, several are ornamentals, and cascara sagrada is a purgative used by Indians of the Pacific Northwest.

Selected Genera:

Rhamnus (includ. *Frangula*) (155)- buckthorn, cascara

Ceanothus (80)- California lilac, New Jersey tea

Paliurus (8)- Christ thorn

Zizyphus (100)- jujube; an edible fruit

Berchemia (22)- supple jack; a stout vine

Colletia (17)- anchor plant; ornamental spiny shrubs

Recognition Characters:

Trees and shrubs with perigynous flowers and stamens opposite the petals.

$$\underline{K}\,^5\,C\,^5\,\underline{A}\,^5\,\underline{G}\,^{(3)}$$

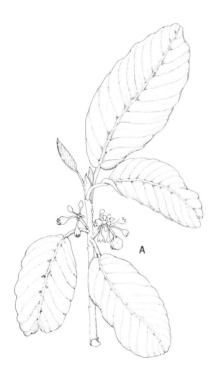
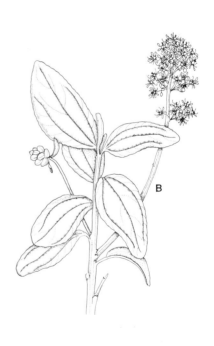

Fig. 66. **Rhamnaceae.** a, *Rhamnus;* b, *Ceanothus.*

VITACEAE
Grape Family

Mostly climbing shrubs, rarely erect shrubs or trees. Stems tendril-bearing, these representing the main axis, the lateral branch continuing as what appears to be the main axis. Leaves alternate, simple or variously compound, often dotted, stipules present or absent. Flowers bisexual or unisexual (the species usually monoecious), actinomorphic, borne in cymes opposite the leaves. Calyx of 4-5 small, cup-like sepals. Corolla of 4 or 5 petals, often united at the tips and falling with the opening of the bud. Androecium of 4 or 5 stamens, these opposite the petals. Gynoecium of 2 [3-6] united carpels, placentation axile, ovary superior. Fruit a berry.

12/700; tropical and subtropical. The four genera listed below are native to the U.S. Of economic importance because of *Vitis vinifera*, whose fruits are used in wine making, raisins, and grape sugar. A few are widely cultivated ornamentals. The family name has been conserved over the Ampelidaceae and Vitidaceae.

Selected Genera:

Vitis (50+)- grape; *V. vinifera* is the European grape of commerce

Cissus (350+)- possum grape; some are succulent ornamentals

Parthenocissus (15)- Virginia creeper, a common ornamental

Ampelopsis (20)- peppervine

Recognition Characters:

Tendril-bearing climbing shrubs with inflorescences oppsoite the leaves, stamens opposite the petals, fruit a berry.

K 4-5 C 4-5 A 4-5 G̲ (2)

HIPPOCASTANACEAE
Buckeye Family

Trees or shrubs with large, glutinous ter-minal buds. Leaves opposite, palmately compound. Flowers bisexual, zygomorphic, with 5 connate sepals, 5 or 4 distinct petals, 8-5 stamens, and 3 united carpels. The fruit is a leathery 1-seeded capsule.

2/15; mostly in North and South America. A few species of *Aesculus* are indigenous to the U.S. The family is of limited economic importance as a source of ornamentals.

Genera:

Aesculus (13)- buckeye, horsechestnut

Billia (2)- South America to Mexico

Recognition Characters:

Woody plants with opposite palmately compound leaves. Flowers in conspicuous inflorescences. Fruit a leathery 1-seeded capsule.

K (5) Cz 5 or 4 A 8-5 G̲ (3)

ACERACEAE
Maple Family

Trees and shrubs. Sap often milky. Leaves opposite, simple, (pinnately compound in *Acer negundo*), palmately lobed and veined, exstipulate. Flowers unisexual (the species monoecious, dioecious, or polygamous) or bisexual, actinomorphic, in racemose, corymbose, or fasciculate inflorescences. Calyx of 4 or 5 connate or distinct sepals. Corolla of 4 or 5 [0] distinct petals, staminal disc usually present. Androecium of 4-10 stamens. Gynoecium of 2 united carpels, laterally compressed, 2-locules, placentation axile, ovary superior. Fruit variously interpreted as either two samaras, separating at maturity, or a samaroid schizocarp, separating into 2 winged mericarps at maturity.

2/200; temperate northern hemisphere. About 15 species of *Acer* are native to the U.S. The family is of some economic importance as a source of timber, ornamentals, and sugar.

Genera:

Acer (including*Negundo*) (200)- maple, box elder

Dipteronia (2)- native to China; the fruits winged all around

Recognition Characters:

Trees and shrubs with opposite palmately lobed and veined leaves (compound in *Acer negundo*). Fruits paired winged schizocarps.

Fig. 67. **Hippocastanaceae.** a, *Aesculus.* **Aceraceae.** b-d, *Acer* flowering branch, schizocarp, and flower.

ANACARDIACEAE
Cashew Family

Trees or shrubs, rarely vines, often with resinous bark. Leaves alternate, simple or pinnately compound, exstipulate. Flowers bisexual or unisexual (the species polygamous), small, actinomorphic, typically 5-merous, borne in panicles. Calyx of 5 [3] basally connate sepals. Corolla of 5 [0, 3] distinct or basally connate petals. Androecium of 10 [5] stamens, those associated with a staminal disc. Gynoecium of 3 united carpels, unilocular by abortion, placentation axile, ovary superior. Fruit a drupe.

79/600; primarily tropical, but extending into the Mediterranean, Eurasia, and North America. About 7 genera are indigenous to the U.S. Of considerable economic importance, although not widely known in the U.S. Products include resins, oils, lacquers, edible fruits, ornamentals, and tannic acid. Although many members of the Anacardiaceae are edible, care should be exercised in dealing with the family. The resinous oils produce extreme dermatitis in sensitive individuals.

Selected Genera:

A. Leaves simple; carpels 5 and free or 1

Tribe: Anacardieae

Anacardium (15)- *A. occidentale* is the cashew

Mangifera (40)- *M. indica* is the mango

A. Leaves usually compound; carpels 3 and united, rarely 1

Tribe: Spondieae (ovule 1 in each carpel)

Spondias (6)- mombin, a tropical fruit

Harpephyllum (1)- *H. caffrum* is the kaffir plum

Tribe: Rhoideae (ovule 1 per gynoecium)

Schinus (30)- pepper tree, an ornamental

Cotinus (3)- smoke tree, an ornamental

Metopium (2)- *M. toxiferum* is the poison wood

Rhus (60)- sumac, lemonade berry

Toxicodendron (30)- poison ivy, poison oak, poison sumac

Pistacia (9)- *P. vera* is the pistachio nut

Recognition Characters:

(Ours) trees and shrubs with pinnately compound leaves (often 3 leaflets), resin ducts, 5-merous flowers, staminal disc and unilocular ovary.

$$K \underline{5} C 5 A 10 \underline{G} (3)$$

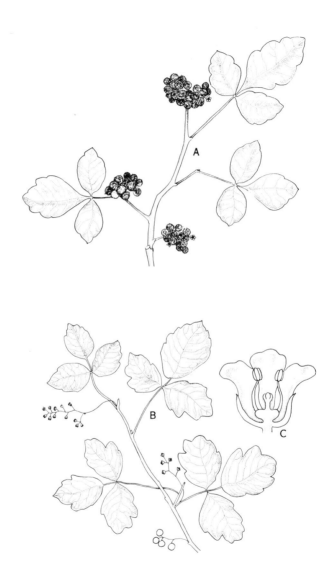

Fig. 68. **Anacardiaceae.** a, *Rhus;* b & c, *Toxicodendron* plant and flower in longitudinal section.

RUTACEAE
Citrus Family

Shrubs and trees, often xerophytic. Leaves alternate or opposite, simple, pinnately or palmately compound, exstipulate, glandular-punctate, often strongly aromatic. Flowers usually bisexual (if unisexual, the species dioecious), actinomorphic or zygomorphic, often in cymose inflorescences. Calyx of 5-4 basally connate or distinct sepals. Corolla of 5-4 [0] distinct petals. Androecium of 10 or 8 stamens, attached at the base or rim of a staminal disc, the outer stamens opposite the petals. Gynoecium of 5 or 4 [1-3-many] united carpels, as many locules, placentation axile, ovary superior, often deeply lobed. Fruit a hesperidium, drupe, berry, samara, or schizocarp.

150/1600; tropical and temperate regions, especially in Australia and South Africa. Eight genera are native to the U.S. Of major economic importance because of the citrus fruits and ornamentals. As treated here, the family includes the Aurantiaceae.

Selected Genera:

Ruta (60)- rue; cultivated for medicinal uses

Zanthoxylum (15)- prickly ash

Dictamnus (2)- dittany; ornamental

Phellodendron (10)- cork oak

Ptelea (3)- hop tree, wafer ash

Citrus (60)- lemon, lime, orange, grapefruit

Fortunella (6)- kumquat

Recognition Characters:

Aromatic shrubs and trees with glandular punctate leaves, and a deeply-lobed ovary.

$$K \ 5\text{-}4 \ C \ 5\text{-}4 \ A \ 10 \text{ or } 8 \ \underline{G} \ (5\text{-}4)$$

ZYGOPHYLLACEAE
Caltrop Family

Herbs and shrubs, rarely trees. Xerophytic and halophytic. Leaves usually opposite, pinnately compound, stipules present and often modified. Flowers usually bisexual, actinomorphic, in cymose inflorescences. Calyx of 5 (rarely 4) separate sepals. Corolla of 5 (rarely 4) separate petals. Androecium of 5, 10 or 15 stamens, the filaments often appendaged. Gynoecium of 5 [2-12] united carpels, the ovary superior. Fruit a loculicidal or septicidal capsule, or drupe-like.

30/250; primarily tropical and subtropical. Six genera are native to the U.S., mostly in the southern half. Guaiacum wood and lignum vitae are the only important economic products.

Selected Genera:

Guaiacum (6)- *G. officinale* is lignum vitae, the densest and hardest wood

Larrea (5)- *L. tridentata* is the creosote bush

Tribulus (20)- *T. terrestris* is the introduced puncture vine

Peganum (5-6)- *P. harmala* is the Syrian rue, whose seeds contain psychoactive alkaloids

Recognition Characters:

Herbs or shrubs with opposite pinnately compound leaves with persistent stipules. Flowers 4 or 5-merous, the stamens often with a scale-like appendage; carpels 5, style 1.

$$K \ 5 \ C \ 5 \ A \ 5,10,15 \ \underline{G} \ (5)$$

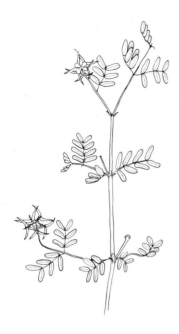

Fig. 69. **Zygophyllaceae.** *Tribulus* with flowers and beaked fruit.

SAPINDACEAE
Soapberry Family

Trees and shrubs, many tendril-bearing lianas. Leaves alternate, pinnately compound or simple, usually exstipulate. Flowers small, functionally unisexual (the species polygamous), zygomorphic or actinomorphic, in racemose or cymose inflorescences. Calyx of 5 distinct sepals. Corolla of 5 [0] petals, often with hairy or scaly glands. Androecium of 5+5 [8, 5, 4 or numerous] stamens, staminal disc present, filaments often hairy. Gynoecium of 3 united carpels, 3-loculed, placentation axile, ovary superior. Fruit a capsule, nut, berry, drupe, schizocarp, or samara. Seeds often with arils.

150/2000; tropical and subtropical. Six genera are indigenous to the U.S., half of these restricted to Florda. Of minor economic importance; the golden raintree and soapberry are cultivated. A few furnish edible fruits; several yield somewhat valuable timber.

Selected Genera:

Sapindus (15)- soapberry, an ornamental tree

Koelreuteria (7)- *K. paniculata* is the golden rain-tree

Cardiospermum (11)- balloon vine; naturalized in the East

Litchi (2)- *L. chinensis* is the lychee, its aril and seed edible

Blighia (6)- *B. sapida* is the akee, a toxic and edible fruit

Dodonaea (60)- hop sage, an ornamental tree

Recognition Characters:

(Ours) trees with pinnately compound leaves, small unisexual flowers with glandular petals, stamens, with hairy filaments, 3-carpellate ovary, and arillate seeds.

$$K \, 5 \, C \, 5 \, A \, 10 \, \underline{G} \, (3)$$

OXALIDACEAE
Oxalis Family

Mostly perennial herbs, rarely trees. Many with fleshy tubers or rhizomes. Leaves alternate, pinnately or palmately compound, or simple by reduction, exstipulate, often with acrid taste. Flowers bisexual, actinomorphic, in umbellate inflorescences. Calyx of 5 distinct sepals. Corolla of 5 free to basally connate petals [0 in the cleistogamous flowers which are sometimes formed]. Androecium of 10 stamens, the outer whorl opposite the petals, monadelphous. Gynoecium of 5 free or united carpels, placentation axile, ovary superior, styles 5 and free. Fruit a loculicidal, elastically explosive capsule; rarely a berry.

8/950; primarily tropical and subtropical.

Oxalis is native to the U.S. Of little economic importance. Several are cultivated; the fruits of carambola *(Averrhoa)* are edible. As treated here, the family includes the Averrhoaceae.

Selected Genera:

Oxalis (850)- sorrel, sheep sorrel, sour grass

Averrhoa (2)- carambola; edible fruits

Recognition Characters:

(Ours) herbs with palmately compound leaves, acrid taste, 5-merous flowers, monadelphous stamens, and 5 styles.

$$K \: 5 \; C \: \underline{5} \; A \: \underline{10} \; \underline{G} \: \underline{5}$$

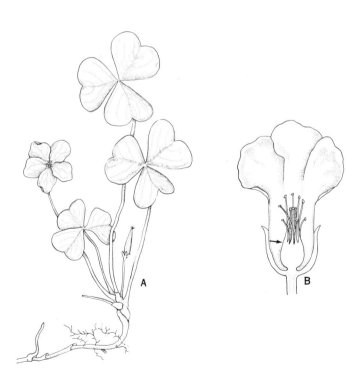

Fig. 70. **Oxalidaceae.** a, *Oxalis;* b, *Oxalis* flower in longitudinal section (arrow indicates monadelphous androecium).

GERANIACEAE
Geranium Family

Herbs, subshrubs, rarely tree-like. Leaves alternate or opposite; palmately or pinnately lobed, dissected or compound; venation often palmate, stipulate. Flowers bisexual, actinomorphic or slightly zygomorphic, cymose or in umbels. Calyx of 5 free or half-connate sepals. Corolla of 5 free petals, often alternating with nectar glands. Androecium of 5, 10, or 15 stamens, sometimes the filaments ± connate at the base. Gynoecium of 5 [3] united carpels, as many locules and styles, placentation axile, ovary superior. Fruit a schizocarp, the one-several seeded mericarps splitting away from and rolling or spiralling up on a central beak.

11/780; cosmopolitan, particularly well represented in the north temperate zone and in South Africa. *Geranium* and *Erodium* are native to the U.S. Of no real economic importance, except as a source of ornamentals, primarily from the cultivated geranium.

Selected Genera:

Geranium (375)- crane's bill; leaves palmate

Erodium (75)- stork's bill; leaves pinnate

Pelargonium (250)- geranium; many cultivars

Recognition Characters:

(Ours) herbs with palmately or pinnately lobed to compound leaves. Flowers 5-merous, the fruits elastically dehiscent schizocarps which curl up on a beak.

$$K^5 C^5 A^{10} \underline{G}^{(5)}$$

Fig. 71. **Geraniaceae.** a, *Erodium;* b & c, *Geranium* fruit before and after dehiscence (arrow indicates position of seeds); d, *Geranium.*

BALSAMINACEAE
Touch-Me-Not Family

Succulent herbs with translucent, watery stems. Leaves alternate or opposite, exstipulate. Flowers bisexual, zygomorphic, solitary to subumbellate. Flowers with 3[5] unequal sepals, the lowermost forming a tubular spur; 5 petals, the lateral pair united; 5 stamens, connate above, forming a cap over the gynoecium; 5 carpels, 5 locules, the ovary superior. Fruit an explosive capsule (hence the common name of the family) or rarely a berry.

4/500-600; primarily Eurasian, North American, and African. *Impatiens* is native to the eastern U.S. The family is of no direct economic importance.

Selected Genus:

Impatiens (419)- jewel weed, balsam, touch-me-not

Recognition Characters:

Herbs with translucent stems. Flowers 5-merous, spurred. Stamens forming a cap over the gynoecium. Fruit an explosive capsule.

$$K_z^{2+1} C_z^{1+(2)+2} \widehat{A^5} \underline{G}^{(5)}$$

Fig. 72. **Balsaminaceae**. Plant and flowers of *Impatiens*.

LINACEAE
Flax Family

Herbs, sometimes shrubs, rarely trees. Leaves alternate or opposite, simple, entire, stipules usually present. Flowers bisexual, actinomorphic, 5-merous, in cymose inflorescences. Calyx of 5 distinct sepals. Corolla of 5 free petals. Androecium of 5, 10, or more stamens, monadelphous. Gynoecium of 5 [3-4] united carpels, placentation axile, ovary superior. Fruit a septicidal capsule or drupe.

12/290; cosmopolitan, but well represented in the temperate regions. *Linum* and *Hesperolinon* are native to the U.S. Of some economic importance because of flax fibers, linseed oil, and ornamentals.

Selected Genera:

Linum (230)- flax, the source of linen and linseed oil

Hesperolinon (12)- a *Linum* segregate of the Pacific states.

Recognition Characters:

(Ours) herbs with 5-merous flowers, often clawed petals, monadelphous stamens, and a septicidal capsule.

$$K 5 \quad C 5 \quad A 5 \text{ or } 10 \quad \underline{G} (5)$$

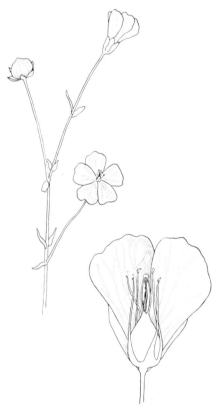

Fig. 73. **Linaceae**. Plant and longitudinal section of *Linum* flower.

POLYGALACEAE
Milkwort Family

Herbs, shrubs, climbers, or small trees. Leaves alternate, simple, usually exstipulate. Flowers bisexual, zygomorphic, subtended by a bract and pair of bracteoles, borne in racemes, spikes, or panicles. Flowers usually with 5 distinct sepals, the inner two often enlarged and petaloid. Corolla usually of 3 petals, these joined to the androecial column, the lower petal often fringed. Androecium of 8 (rarely fewer) stamens, monadelphous, the tube split open. Gynoecium of 2 (rarely 5) united carpels, the ovary superior. Fruit a capsule, nut or drupe.

13/800; cosmopolitan, except for New Zealand, the Arctic, and Polynesia. The genera below are native to the U.S. A few species are cultivated as ornamentals. The flower is somewhat reminiscent of a papilionoid legume, with which it is often confused by the beginning student.

Selected Genera:

Polygala (500)- milkwort

Monnina (80)- found in New Mexico and Arizona

Recognition Characters:

Shrubs and herbs with 5-merous flowers, the two inner sepals enlarged and petaloid; the corolla often reduced to three petals, (one often fringed), these joined to a staminal tube; the gynoecium bicarpellate.

$$Kz^{3+2} \; C^3 \; A\underline{8} \; \underline{G}^{(2)}$$

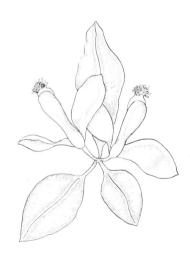

Fig. 74. **Polygalaceae.** *Polygala* flower.

ARALIACEAE
Ginseng Family

Mostly woody plants, sometimes lianas. Stems solid, often prickly. Leaves usually alternate, palmately or pinnately compound, with stellate hairs, stipules present, often adnate to the petiole base. Flowers unisexual (the species dioecious or polygamous) or bisexual, small greenish, actinomorphic; in heads or umbels, these sometimes racemose. Calyx of 5 small sepals. Corolla of [4] 5 [10] petals. Androecium of 5 [3- many] stamens, arising from a staminal disc. Gynoecium of 5 [1- many] united carpels, placentation axile, ovary inferior. Fruit a berry or drupe.

70/700; primarily tropical, particularly well represented in Indomalaysia and tropical America. *Aralia*, *Oplopanax*, and *Panax* are native to the U.S. Of limited economic importance; several are ornamentals and there is trade in ginseng root, the famous medicinal panacea.

Selected Genera:

Aralia (30)- hercules club, angelica tree

Fatsia (1)- *F. japonica* is commonly cultivated

Tetrapanax (1)- *T. papyriferus* is the rice-paper plant

Schefflera (150)- octopus tree; ornamental

Hedera (6)- English ivy; common ornamental

Panax (6)-*P. quinquefolius* is the ginseng of the U.S.

Recognition Characters:

Shrubs and small trees with racemose umbels of 5-merous, small, epigynous flowers. Fruit a berry.

$$K\ 5\ C\ 5\ A\ 5\ \overline{G}\ (5)$$

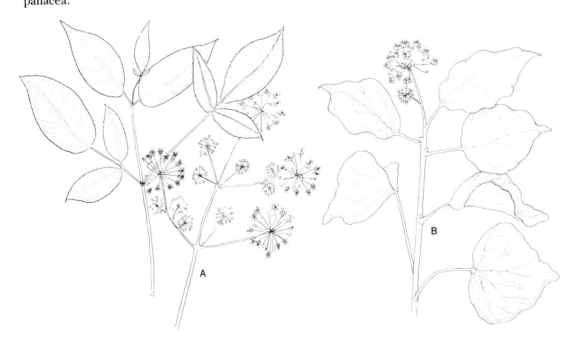

Fig. 75. **Araliaceae.** a, *Aralia;* b, *Hedera.*

UMBELLIFERAE (= APIACEAE)
Parsley Family

Biennial or perennial herbs, sometimes woody, rarely trees. Often aromatic. Stems usually stout, furrowed, with hollow internodes. Leaves alternate, usually compound, with sheathing leaf bases. Flowers usually bisexual, actinomorphic, 5-merous, epigynous, usually in compound umbels, less frequently in simple umbels or heads. Calyx of 5 small distinct sepals. Corolla of 5 [0] distinct petals, often yellow or white. Androecium of 5 stamens. Gynoecium of 2 united carpels, 2-loculed, placentation axile, ovary inferior. Above the ovary is the **stylopodium,** the fleshy fused bases of the 2 styles. Fruit a schizocarp, separating into 2 one-seeded mericarps, suspended on a **carpophore,** the dry wiry remnants of vascular strands.

Much of the taxonomy of the Umbelliferae depends upon details of the schizocarp. The two mericarps that form the schizocarp come together along a suture called a **commissure.** The outer wall of the mericarp usually has 5 **primary ridges;** 2 **lateral ridges** running along the edges bordering the commissure, 2 **intermediate** ridges above these, and a central dorsal ridge. Four **secondary ridges** may occur among the 5 primary ridges. In the valleys between the ridges may be found oil passages called **vittae.**

300/3000; primarily of the temperate northern hemisphere. Tropical representatives are usually mountainous. About ¼ of the genera and 10 percent of the species are native to the U.S.

Of considerable economic importance because of numerous food plants, condiments, ornamentals, and poisonous species. Generally speaking the Umbelliferae should not be eaten unless identification is certain. *Conium maculatum* and *Cicuta* spp. are lethal. Several species, including the wild parsnip (*Pastinaca sativa*) can cause severe dermatitis in sensitive individuals. An unacceptable alternative family name, Ammiaceae, often appears in the literature. As treated here, the family includes the Hydrocotylaceae, Angelicaceae, Daucaceae, and Umbellaceae.

Selected Genera:

A. Stipules present

Subfamily: Hydrocotyloideae

Hydrocotyle (78)- aquatics with peltate leaves

A. Stipules absent

Subfamily: Saniculoideae (styles long; stigmas capitate)

Sanicula (40)- sanicle, snakeroot

Eryngium (220)- eryngo, rattlesnake master

Subfamily: Apioideae (styles short; stigmas not capitate)

Osmorhiza (11)- sweet cicely

Coriandrum (2)- *C. sativum* is coriander

Conium (2)- poison hemlock; lethal if consumed

Apium (20)- *A. graveolens* is celery

Petroselinum (4)- *P. crispum* is parsley

Cicuta (7)- water hemlock; violently lethal

Carum (25)- caraway

Pimpinella (150)- *P. anisum* is anise

Foeniculum (2-3)- *F. vulgare* is fennel

Anethum (2)- *A. graveolens* is dill

Angelica (50)- large perennials; ornamentals

Ferula (50)- *F. asafoetida* yields asafetida

Pastinaca (14)- *P. sativa* is the parsnip

Heracleum (60)- *H. lanatum* is the cow parsnip

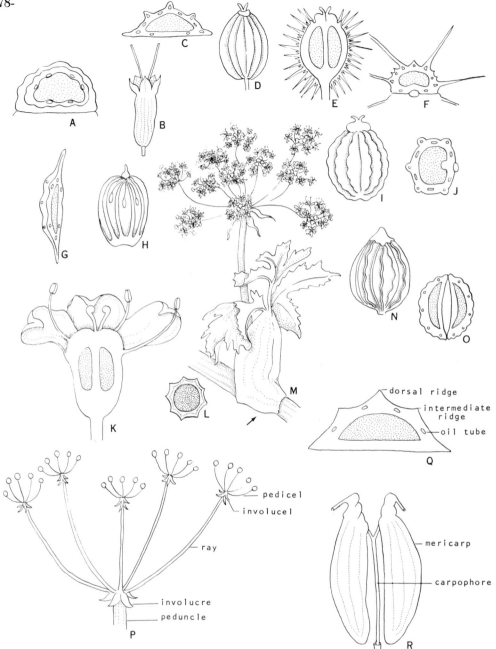

Fig. 76. **Umbelliferae.** a & b, fruit of *Oenanthe;* c & d, fruit of *Levisticum;* e & f, fruit of *Daucus;* g & h, fruit of *Heracleum;* i & j, fruit of *Conium;* k, generalized flower; l, stem in x-s at a node; m, *Heracleum* leaf and inflorescence (arrow indicates sheathing base); n & o, fruit of *Coriandrum;* p, generalized compound umbel; q, mericarp of a fruit; r, schizocarp, the mericarps having separated. (Fruit drawings after Melchior and Werdermann, 1964).

Daucus (60)- *D. carota* is the carrot

Recognition Characters:

Aromatic herbs with hollow, furrowed stems, compound leaves with sheathing leaf bases, 5-merous often white or yellow flowers in compound umbels, and fruit a schizocarp.

$$K \, 5 \, C \, 5 \, A \, 5 \, \overline{G}(2)$$

Magnoliopsid

Order: Family	Common Name	Genera/Species	Distribution	Form	K
Gentianales					
Loganiaceae	Logania	32/800	Pantrop.	HST	(4-5)
Gentianaceae	Gentian	70/1100	Cosmo.	H	4
Apocynaceae	Dogbane	200/2000	Cosmo.	SVH	(5)
Asclepiadaceae	Milkweed	250/2000	Trop. & subtrop.	HS	5
Polemoniales					
Solanaceae	Nightshade	85/2300	Trop. & subtrop.	HSVT	(5)
Convolvulaceae	Morning glory	50/1650	Trop. & temp.	HSV	5
Cuscutaceae	Dodder	1/170	Cosmo.	H	4-5
Menyanthaceae	Bog bean	5/33	N. & S. temp.	H	5
Polemoniaceae	Phlox	18/320	Chiefly N. Am.	H	(5)
Hydrophyllaceae	Waterleaf	20/270	Cosmo.	H	(5)
Lennoaceae	Lennoa	3/5	SW U.S. & Mex.	H	5-10
Lamiales					
Boraginaceae	Borage	100/2000	Trop. & temp.	H	$\underline{5}$
Callitrichaceae	Water starwort	1/25	Cosmo.	H	0
Verbenaceae	Verbena	75/3000	Trop. & subtrop.	HST	(5)
Phrymaceae	Phryma	1/2	E. As. & E. N. Am.	H	(5)z
Labiatae	Mint	180/3500	Cosmo.	HS	(5)
Plantaginales					
Plantaginaceae	Plantain	3/270	Cosmo.	H	(4)

Asteridae

C	A	G	Fruit Type(s)	Miscellaneous Comments
(4-5)	4-5	(2)	cap	Four genera native to U.S.
(4-5)	4-5	(2)	cap, ber	Leaves opposite; gynoecium unilocular
(5)	5	2	fol, ber, cap	Often with milky sap
(5)	5	2	fol	Often with milky sap
(5)	5	(2)	ber, cap	Family of great economic importance
(5)	5	(2)	cap, ber, nut	Often with milky sap
(4-5)	4-5	(2)	cap	Twining parasites
(5)	5	(2)	cap	Aquatic plants
(5)	5	(3)	cap	Mostly in western U.S.
(5)	5	(2)	cap	Mostly in western U.S.
(5-8)	5-10	(6-14)	cap	Fleshy, achlorophyllous root parasites
(5)	5	(2)	nutl, ach, drp	Ovary 4-lobed
0	1	(2)	schizo	Aquatics; flowers unisexual
(5)z	2+2	(2)	drp	Fourteen genera native to U.S.
(5)z	2+2	(2)		Separated from Verbenaceae by ovule position
(5)z	2 or 2+2	(2)	nutl, drp	Ovary 4-lobed
(4)	4	(2)	cap, nut	*Plantago* only widespread representative

Scrophulariales

Buddlejaceae	Buddleja	6-10/150	Trop. & temp.	TS	(4)
Oleaceae	Olive	29/600	Cosmo.	ST.	(4)
Scrophulariaceae	Figwort	220/3000	Cosmo.	HS	(5)
Gesneriaceae	Gesneria	140/1800	Trop. & subtrop.	H	5
Orobanchaceae	Broomrape	13/180	N. temp.	H	(2-5)
Bignoniaceae	Bignonia	120/800	Trop.	TS	(5)
Acanthaceae	Acanthus	250/2600	Pantrop.	HS	(4-5)
Pedaliaceae	Pedalium	16/55	Chiefly OW trop.	H	(5)
Lentibulariaceae	Bladderwort	5/300	Cosmo.	H	2-5

Campanulales

Campanulaceae	Harebell	70/2000	Widespread	HS	5

Rubiales

Rubiaceae	Madder	500/6-7000	Pantrop.	TSH	4-5

Dipsacales

Caprifoliaceae	Honeysuckle	15/400	N. temp.	S	5
Adoxaceae	Adoxa	1/1	N. Am.	H	3+5or4
Valerianaceae	Valerian	13/400	N. temp. & Andes	H	0
Dipsacaceae	Teasel	10/270	OW	H	(5)

Asterales

Compositae	Sunflower	920/19,000	Cosmo.	HST	x

(4)	4	(2)	cap, drp, ber	Differ from Loganiaceae in xylem structure
(4)	2	(2)	ber, drp, schizo, cap	One of the few woody families with two stamens
(5)z	[2]2+2[5]	(2)	cap, ber	Temperate counterpart of several tropical families
(5)z	2+2	(2)	cap	Differ from figworts in unilocular ovary
(5)z	2+2	(2)	cap	Achlorophyllous root parasites
(5)z	2+2	(2)	cap	Differ from figworts in absence of endosperm
(4-5)z	2+2	(2)	cap	Flowers often subtended by showy bracts
(5)z	2+2	(2)	cap, nut	Four-loculed, unlike other families of order
(5)z	2	(2)	cap	Note the reduced stamen number
(5)or(5)z	5	$\overline{(2,3,5)}$	cap, ber	Actinomorphic or zygomorphic (= Lobeliaceae)
(4-5)	4-5	$\overline{(2)}$	cap, ber	Stipules often leaf-like
(5)or(5)z	4-5	$\overline{(2 or 8)}$	ber, drp, cap	Exstipulate; easily confused with Rubiaceae
	5-4	$\overline{(3-5)}$	drp	In northern part of U.S.
(5)z	1-4	$\overline{(3)}$	ach	Flowers spurred or saccate
(4-5)	4	$\overline{(2)}$	ach	None is native, but *Dipsacus* widely naturalized
(5)or(5)z	5	$\overline{(2)}$	ach	Capitulum is typical inflorescence type

The Asteridae are a subclass of nine orders, 43 families, and about 56,000 species. It is the second largest subclass of the Magnoliopsida. Most members of the Asteridae are sympetalous, have as many or fewer stamens than corolla lobes, and are bicarpellate. It is the only subclass not derived directly from the Magnoliidae.

The orders may be characterized as follows. Plants of the Gentianales have simple opposite leaves; hypogynous regular flowers whose lobes are contorted in bud; and a bicarpellate gynoecium. The Polemoniales are much like the Gentianales, except for their alternate leaves. The Lamiales are distinguished by their bicarpellate gynoecium, each carpel with two ovules and divided by a partition or appearing quite separate. The Plantaginales have reduced, 4 - merous, anemophilous flowers. The Scrophulariales are exstipulate plants with hypogynous often zygomorphic flowers. The Campanulales are mostly herbaceous plants with an inferior ovary and ± connivent to connate anthers. The Rubiales are mostly herbs in our range with oppposite or whorled leaves, as inferior ovary and ± regular flowers. The Dipsacales are herbs or shrubs with opposite leaves, zygomorphic flowers, epipetalous stamens, and an inferior ovary. The characters of the Asterales are those of its only family, the Compositae.

GENTIANACEAE
Gentian Family

Annual or perennial herbs, rarely woody. Leaves opposite, entire, estipulate, often connate at the base. Flowers usually bisexual, actinomorphic, usually in cymes. Calyx of 4 (5) free sepals. Corolla of 4 or 5 united petals; campanulate, funnelform, or sometimes salverform. Androecium of as many stamens as petals, epipetalous. Gynoecium of 2 united carpels, unilocular, with a glandular disc at its base, placentation parietal, ovary superior. Fruit a septicidal capsule; rarely a berry.

70/1100; widespread in temperate and sub-tropical areas. Thirteen genera are native to the U.S. Several are cultivated as ornamentals. As treated here, the family excludes the Menyanthaceae.

Selected Genera:

Gentiana (200+)- gentian

Centaurium (40)- centaury

Sabbatia (15)- marsh pink

Swertia (100)- felwort

Eustoma (3)- prairie gentian; catch fly gentian

Frasera (15)- green gentian

Recognition Characters:

Herbs with opposite, exstipulate, basally connate leaves, 4 or 5-merous flowers with a superior, bicarpellate, unilocular gynoecium.

$$K^4 \ C^{(4\text{-}5)} \ A^{4\text{-}5} \ \underline{G}^{(2)}$$

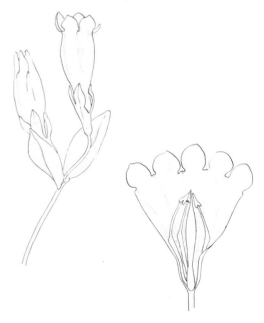

Fig. 77. **Gentianaceae.** a, flowers of *Gentiana*; b, longitudinal section of flower.

APOCYNACEAE
Dogbane Family

Trees, shrubs, or rarely perennial herbs. Usually with milky latex. Leaves opposite or verticillate, simple, entire, exstipulate. Flowers bisexual, usually actinomorphic, hypogynous, in racemose or cymose inflorescences. Calyx of 5[4] sepals, often glandular inside. Corolla of 5 united petals, tubular. Androecium of 5[4] stamens, epipetalous, anthers often sagittate, free or connivent about the stigma, but **not** usually adherent to it, pollen granular or in tetrads. Gynoecium of 2 carpels, these free or united by the thickened styles; the comparatively massive distal stigmatic area referred to as the **clavuncle;** each carpel with marginal placentation and superior to partly inferior. Fruit 2 follicles, a berry or capsule. Seeds sometimes comose.

200/2000; cosmopolitan, particularly pantropical. Eleven genera are native to the U.S., about half of them in Florida. The source of several fine ornamentals; most members of the family are poisonous. This family name is conserved over the Plumeriaceae.

Selected Genera:

Subfamily: Plumerioideae (stamens free from clavuncle; seeds hairless)

Ervatamia (100)- crape jasmine

Catharanthus (6)-*C. roseus* is the periwinkle

Vinca (6)- periwinkle, grave myrtle

Rauvolfia (90)-Indian snakeroot; source of reserpine

Allamanda (18)- cups of gold

Plumeria (7)- frangipani

Subfamily: Apocynoideae (stamens joined to clavuncle; seeds hairy)

Apocynum (7)- dogbane

Nerium (3)-*N. oleander* is the oleander, a toxic shrub

Recognition Characters:

Herbs, shrubs, or small trees with opposite or whorled leaves, milky sap, 5-merous flowers, gynoecium of 2± separate carpels, fruit often a follicle filled with comose seeds. The dogbane family is most easily confused with the Asclepiadaceae. The Apocynaceae lack the well-developed corona, pollinia, and translator - corpusculum unit of the milkweed family.

$$K^{(5)} \ C^{(5)} \ A^5 \ \underline{G}^{\overparen{2}}$$

Fig. 78. **Apocynaceae.** a, *Nerium*, dehisced follicle; b, seed; c, *Apocynum;* d, *Allamanda;* e, *Vinca* flower in longitudinal section (arrow indicates clavuncle).

ASCLEPIADACEAE
Milkweed Family

Perennial herbs, erect or twining shrubs, rarely trees, sometimes fleshy and cactoid. Milky latex usually present. Leaves opposite or whorled, sometimes alternate, simple, entire, linear to obovate, exstipulate, in cactoid forms reduced or obsolete. Flowers bisexual, actinomorphic, 5-merous, hypogynous, in cymose, often umbellate inflorescences. Calyx of 5 distinct or basally connate sepals, reflexed. Corolla of 5 united petals. A corona may be present, its structure variable, but typically consisting of 5 erect, often brightly colored elements sometimes mistaken for petals. Each component is called a **hood.** Associated with each of the 5 hoods may be a beak-like structure pointing toward the center of the flower. This is the **crest** or **beak.** Within the corona is an androecium of 5 stamens, these adherent in a ring to the surface of the stigmatic area of the gynoecium. Together the androecium and gynoecium form a compound structure, the **gynostegium.** Adjacent ánthers are joined together by 2 arms **(translators** or **connectives)** to a central body, the **corpusculum** or **gland.** The translators may be formed by secretions from the stigma. The half - anthers of adjacent stamens are, therefore, connected to the corpusculum by the translators. Each half - anther contains pollen grains united into waxy masses, the entire mass being termed a **pollinium.** The pollinia are usually situated behind the hood elements, with the translator arms and corpusculum often visible between the corona elements. During pollination an insect removes a pair of pollinia from adjacent stamens and the associated structures by snagging them on some tiny spur on its leg as it visits the milkweed flower. For cross-pollination to be effected, the pollinia must be reinserted in a precise fashion in vacant slots behind the hood elements of another flower. If the pollinia are reversed, the pollen masses germinate with the pollen tubes growing away from the stigmatic survace. Perhaps this partially explains the presence of so few fruits on milkweed plants. The gynoecium is composed of 2 carpels, free through much of their length; ovary superior, placentation marginal. Fruit a pair of follicles, one from each of the essentially distinct carpels. Seeds typically with a coma of very silky hairs.

250/2000, primarily tropical and subtropical. Generic deliminations are quite controversial. Following a conservative view, there are 5 genera represented in the U.S. The family is of moderate economic importance as a source of ornamentals, latex, fibers, poisonous plants, and a few food plants. Several species of *Asclepias* are poisonous to humans and livestock. The ceylon milk plant (*Gymnema lactiferum*) is edible. Generally, however, plants of this family and all plants with milky latex should be regarded as inedible and potential sources of severe dermatitis. As treated here, the family includes the Stapeliaceae and Periplocaceae.

Selected Genera:

Asclepias (including *Asclepiodora* and *Acerates*) (80-100)- milkweed

Hoya (100+)- carnosa wax plant

Stapelia (100)- carrion flower; succulent ornamentals

Cryptostegia (2)- a latex plant from Madagascar

Recognition Characters:

(Ours) herbs with opposite or whorled leaves, milky latex, 5-merous flowers, corona, pollinia, translators, corpuscula, and follicles of comose seeds.

$$K^5 \ C^{(5)} \ A^{\widehat{5}} \ \underline{G}^{\widehat{2}}$$

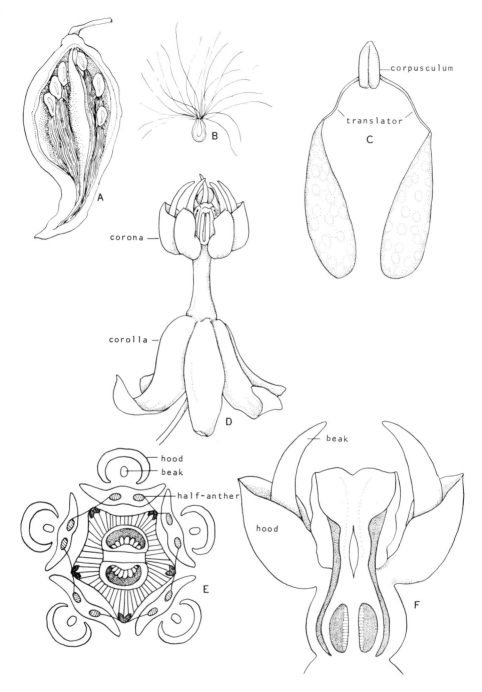

Fig. 79. **Asclepiadaceae.** a, follicle after dehiscence; b, seed; c, pollinia, corpusculum, and translator; d, flower; e, flower in x-s; f, flower in longitudinal section. (Drawings e & f after Melchior and Werdermann, 1964).

SOLANACEAE
Nightshade Family

Herbs, shrubs, lianas, and trees. Leaves alternate, simple, and exstipulate. Flowers bisexual, actinomorphic or rarely zygomorphic, 5-merous, hypogynous, in cymose inflorescences. Calyx of [4] 5 [6] persistent, connate sepals, sometimes enlarging in fruit. Corolla of 5 united petals, usually regular, rotate to tubular, folded and convolute. Androecium of 5 [4, 2+2, or 2, particularly in zygomorphic flowers] stamens, epipetalous. Gynoecium of 2 united carpels, 2-loculed, placentation axile, ovules numerous, ovary superior. Fruit a berry or septicidal capsule.

85/2300, tropical and subtropical; the center of distribution is Central and South America. About 13 genera are indigenous to the U.S. Of great economic importance because of numerous food plants, ornamentals, some medicinal plants, and several poisonous ones. Many species of *Solanum* are referred to as nightshades, an allegorical reference to their poisonous properties.

Selected Genera:

A. Embryo definitely curved; all 5 stamens fertile, ± equal in length

Tribe: Solaneae (ovary 2-loculed)

Atropa (5)- *A. belladonna* is belladona

Hyoscyamus (15)- *H. niger* is henbane

Physalis (110)- groundcherry, husk tomato

Capsicum (35)- chili, tabasco, bell pepper

Solanum (1500)- nightshade, Irish potato, eggplant

Lycopersicon (10)- *L. esculentum* is the tomato

Mandragora (4)- *M. officinarum* is the mandrake of medicine

Lycium (110)- matrimony vine, box thorn

Tribe: Datureae (ovary 4-loculed)

Datura (18)- jimson weed, sacred datura

A. Embryo straight or slightly curved; 2, 4 or 5 stamens fertile

Tribe: Cestreae (all 5 stamens fertile)

Nicotiana (100)- *N. tabacum* is tobacco

Petunia (14)- petunia

Tribe: Salpiglossideae (2 or 4 stamens fertile)

Salpiglossis (5)- painted tongue, an ornamental

Schizanthus (15)- butterfly flower, an ornamental

Recognition Characters:

Herbs or shrubs (ours) with alternate leaves, actinomorphic, 5-merous flowers with 5 stamens, and a berry or capsule fruit. The family is most easily confused with the Scrophulariaceae. Members of the latter family typically have zygomorphic flowers and 4 stamens. Numerous exceptions occur.

$$K^{(5)} \; C^{(5)} \; A^{5} \; \underline{G}^{(2)}$$

-190-

Fig. 80. **Solanaceae.** a-c, *Datura* flower, fruit, and plant; d-g, *Solanum* fruit, plant, and flower (arrow indicates connivent anthers).

CONVOLVULACEAE
Morning Glory Family

Herbs, shrubs, or trees, often climbing. Milky latex often present. Leaves alternate, simple, exstipulate. Flowers bisexual, regular, bracteate, typically 5-merous, hypogynous, in cymes. Calyx of 5 distinct sepals, these sometimes unequal. Corolla of 5 united, plaited petals. Androecium of 5 stamens, epipetalous; nectar disc usually present. Gynoecium of 2 (3-5) united carpels, placentation axile, ovary superior. Fruit a capsule, berry, or nut.

50/1400-1650; tropical and temperate. Nine genera are native to the U.S. Of some economic importance because of the sweet potato, several weeds, and several ornamentals.

Selected Genera:

Convolvulus (250)- bindweed, wild morning glory

Calystegia (25)- bindweed, wild morning glory

Ipomoea (including *Quamoclit*) (400)- morning glory; *I. batatus* is the sweet potato

Dichondra (5)- dichondra; a grass substitute for lawns

Recognition Characters:

(Ours) trailing vines, often with milky latex, 5-merous flowers with a tubular, plaited corolla, 5 epipetalous stamens, and a bicarpellate gynoecium.

$$K^{5} \; C^{(5)} \; A^{5} \; \underline{G}^{(2)}$$

CUSCUTACEAE
Dodder Family

Leafless, rootless, parasitic herbs; chlorophyll almost totally lacking. Stems thread-like, often yellowish. Flowers small, with 4 or 5 distinct sepals, 4 or 5 united petals, 4 or 5 epipetalous stamens, and 2 united carpels. Fruit a dry or fleshy globose capsule.

1/170; cosmopolitan. *Cuscuta* is native to the U.S., where it causes great losses to crop plants, particularly lespedeza and clover. This family was formerly included in the Convolvulaceae.

Genus:

Cuscuta (170)- dodder

Recognition Characters:

Achlorophyllous, leafless, thread-like plants. Flowers small, 4- or 5-parted.

$$K^{4\text{-}5} \; C^{(4\text{-}5)} \; A^{4\text{-}5} \; \underline{G}^{(2)}$$

Fig. 81. **Convolvulaceae.** a & b, *Calystegia.* **Cuscutaceae.** c, *Cuscuta.*

POLEMONIACEAE
Phlox Family

Herbs, rarely shrubs or small trees. Leaves alternate or opposite, entire, divided, or pinnately compound, exstipulate. Flowers bisexual, actinomorphic, rarely zygomorphic, 5-merous, in cymes. Calyx of 5 united sepals. Corolla of 5 united petals, rotate to salverform. Androecium of 5 epipetalous stamens, these sometimes arising at various levels within the tubular corolla. Gynoecium of [2] 3 [5] united carpels, 3-loculed, the placentation axile, the ovary superior, style 1, stigmas 3. Fruit a loculicidal capsule.

18/320; chiefly North American, particularly in the western U.S. About a dozen genera are native to this country. The source of a few ornamentals.

Selected Genera:

Polemonium (23)- Jacob's ladder

Collomia (14)- several species in western U.S.

Phlox (61)- phlox; both native and cultivated species

Gilia (56)- gilia

Navarretia (30)- spiny annuals, mostly in the west

Cobaea (19)- cup and saucer vine

Recognition Characters:

(Ours) herbs and shrubs with 5-merous, synsepalous, sympetalous flowers, the 5 stamens epipetalous, the gynoecium tricarpellate.

$$K^{(5)} \ C^{(5)} \ A^{5} \ \underline{G}^{(3)}$$

Fig. 82. **Polemoniaceae.** a & b, *Ipomopsis;* c, *Polemonium;* d, *Linanthus.*

HYDROPHYLLACEAE
Waterleaf Family

Annual or perennial herbs, rarely shrubby, often bristly, glandular, or scabrid. Stems round. Leaves alternate or opposite, often with basal rosettes, entire to pinnately lobed, exstipulate. Flowers bisexual, actinomorphic, 5-merous, hypogynous, in cymes, particularly coiled ones. Calyx of 5 united sepals. Corolla of 5 united petals, rotate, campanulate, or funnelform. Androecium of 5 epipetalous stamens, arising from the base of the corolla. Gynoecium of 2 united carpels, unilocular, placentation parietal, ovules numerous, ovary usually superior. Fruit a loculicidal capsule.

20/270; cosmopolitan, except for Australia. One center of distribution is the western U.S. Sixteen genera are indigenous to the U.S. A few are cultivated.

Selected Genera:

Hydrophyllum (8)- water leaf
Nemophila (11)- baby blue-eyes
Phacelia (130)- common in the West
Eriodictyon (10)- yerba santa; a shrub

Recognition Characters:

(Ours) usually bristly herbs (*Eriodictyon* a shrub) with coiled cymose inflorescences of 5-merous, synsepalous, sympetalous flowers with a bicarpellate, unilocular gynoecium. This family is most easily confused with the Boraginaceae from which it differs in having numerous ovules on parietal placentae and an unlobed ovary.

$$K^{(5)} \ C^{(5)} \ A^5 \ \underline{G}^{(2)}$$

Fig. 83. Hydrophyllaceae. a, *Hydrophyllum* (note exserted stamens); b, ovary in x-s showing parietal placentation; c, *Nemophila* flowers; d, *Phacelia* (note coiled inflorescence).

BORAGINACEAE
Borage Family

Mostly herbs, sometimes shrubs or climbers. Plants often with scabrous or hispid hairs. Stems round. Leaves usually alternate, simple, entire, exstipulate. Flowers usually bisexual, actinomorphic, 5-merous, hypogynous, in coiled cymes. Calyx of 5 free or basally connate sepals. Corolla of 5 united petals, salverform, funnelform or campanulate. Androecium of 5 epipetalous stamens. Gynoecium of 2 united carpels, typically 4-loculed by a false septum, 4-lobed, each locule one-seeded, placentation axile, ovary superior, style arising from the base of the ovary. Fruit of 4 nutlets or achenes, fewer by abortion; rarely a drupe.

100/2000; tropical and temperate. Twenty-two genera are native to the U.S., many of them in the West. As treated here, the family includes the Ehretiaceae, Cordiaceae, and Heliotropiaceae.

Selected Genera:

Subfamily: Heliotropioideae (style terminal; fruit a drupe)

Heliotropium (220)- heliotrope

Subfamily: Boraginoideae (style gynobasic; fruit of 4 nutlets)

A. Flowers actinomorphic

    B. Base of style ± conical

        Tribe: Cynoglossoideae (tips of achenes not projecting above point of attachment)

        Cynoglossum (68)- hound's tongue

        Tribe: Eritrichieae (tips of achenes projecting above point of attachment)

        Mertensia (40)- blue bells

        Amsinckia (20)- fiddleneck; toxic

        Cryptantha (150)- white forget-me-not

        Plagiobotrys (60)- popcorn flower

        Myosotis (80)- forget-me-not

    B. Base of style flat or slightly convex

        Tribe: Boragineae (achenes with a concave attachment surface)

        Pulmonaria (12)- lungwort

        Symphytum (17)- comfrey, a popular folk remedy

        Borago (3)- borage, talewort

        Tribe: Lithospermeae (achenes with a flat attachment surface)

        Lithospermum (50)- gromwell, hoary puccoon

A. Flowers zygomorphic

        Tribe: Echieae (often included in the Lithospermeae)

        Echium (50)- viper's bugloss, an introduced toxic shrub

Recognition Characters:

Bristly herbs with alternate leaves and round stems with coiled cymes of 5-merous, often blue flowers; gynoecium bicarpellate, ovary 4-lobed, style arising from among the lobes.

$$K^{(5)} \ C^{(5)} \ A^5 \ \underline{G}^{(2)}$$

Fig. 84. **Boraginaceae.** a, *Borago* flowers; b & c, *Amsinckia* inflorescence and longitudinal section of a flower; d, deeply 4-lobed ovary.

VERBENACEAE
Vervain Family

Herbs, shrubs, or trees. Stems often 4-sided. Leaves usually opposite or whorled, simple, exstipulate. Flowers bisexual, typically zygomorphic, 5-merous, hypogynous, often in cymes. Calyx of 5 [4-8] united sepals. Corolla of 5 petals, salverform, sometimes bilabiate. Androecium of 4 didynamous stamens, rarely 5 or 2. Gynoecium of 2 [4 or 5] united carpels, as many or twice as many locules, placentation axile, the ovary superior, style 1, terminal. Fruit a drupe, nutlet, or capsule.

75/3000; tropical and subtropical. Fourteen genera are native to the U.S. Of economic importance because of the highly-prized teak wood and a number of ornamentals. As treated here, the family includes the Avicenniaceae and Stilbeaceae.

Selected Genera:

A. Ovules anatropous (micropyle close to funiculus attachment)

Subfamily: Verbenoideae (inflorescence racemose or spicate)

Verbena (230)- vervain, verbena

Lantana (160)- lantana, a toxic shrub

Phyla (including *Lippia*) (15)- frogfruit

Duranta (25)- pigeon berry

Subfamily: Viticoideae (inflorescence cymose)

Callicarpa (100)- beauty berry

Tectona (4)- *T. grandis* is teak, prized for its wood

Clerodendrum (390)- glory bower

Vitex (270)- chaste tree

A. Ovules orthotropous (erect and straight)

Subfamily: Avicennioideae (= Avicenniaceae of many authors)

Avicennia (11)- black mangrove of Florida and Louisiana coastal swamps

Recognition Characters:

(Ours) herbs and shrubs with opposite leave, 5-merous zygomorphic flowers with a single terminal style. The family is most easily confused with the Boraginaceae, which have alternate leaves and a 4-lobed ovary; and the Labiatae, which have a 4-lobed ovary and a style arising from the base of the ovary.

$$K^{(5)} \; Cz^{(5)} \; A^{2+2} \; \underline{G}^{(2)}$$

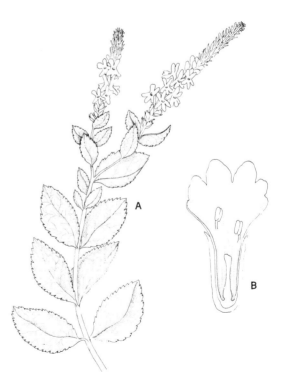

Fig. 85. **Verbenaceae.** a & b, *Verbena* plant and flower in longitudinal section.

LABIATAE (= LAMIACEAE)
Mint Family

Aromatic herbs, sometimes shrubs, rarely trees or lianas. Stems usually 4-sided. Leaves opposite or whorled, usually simple, often hairy, with epidermal glands secreting various volatile oils. Flowers bisexual, usually zygomorphic, 5-merous, in whorls or axillary. Calyx of 5 united sepals, sometimes bilabiate. Corolla or 5 united petals, bilabiate with 2 fused upper petals and 3 fused lower ones. Androecium of 2 or 4 epipetalous stamens, if 4 then usually didynamous; nectar disc often present. Gynoecium of 2 united carpels, 4-lobed, locules 2, often appearing as 4 by secondary constriction of ovary wall, placentation axile (often appearing basal by reduction of ovule number), ovary superior, style 1, arising from among the lobes of the ovary. Fruit usually 4 nutlets, sometimes fewer by abortion; rarely a drupe.

180/3500; cosmopolitan, the chief center of distribution the Mediterranean region. Almost 50 genera occur in the U.S. Of considerable economic importance as a source of numerous ornamentals, aromatic oils used to flavor foods, and a few weedy genera. The family name is conserved over the Menthaceae.

Selected Genera:

A. Style not gynobasic; nutlets with a lateral-ventral attachment and large surface of contact with one another)

    Subfamily: Ajugoideae (upper lip rarely concave; nutlets wrinkled)

        Trichostema (10)- bluecurls, vinegar weed

        Teucrium (100)- germander

        Ajuga (45)- bugle weed

    Subfamily: Rosmarinoideae (upper lip concave; nutlets smooth)

        Rosmarinus (1)- *R. officinalis* is rosemary

A. Style gynobasic; nutlets with basal attachment and small surface of contact with one another)

B.   Seeds transverse

    Subfamily: Scutellarioideae

        Scutellaria (200)- skull cap, a reference to the protuberance on the calyx

B.   Seeds erect

    Subfamily: Lavanduloideae

        Lavandula (26)- lavender

    Subfamily: Lamioideae (= Stachyoideae)

        Mentha (15)- mint, peppermint, spearmint

        Origanum (40)- *O. majorana* is marjoram

        Thymus (35)- *T. vulgaris* is thyme

        Satureja (130)- savory, yerba buena

        Marrubium (30)- *M. vulgare* is hoarhound

        Nepeta (150)- *N. cataria* is catnip

        Glechoma (5)- *G. hederacea* is ground ivy

        Stachys (200)- betony, hedge nettle

        Molucella (2)- *M. laevis* is bells - of Ireland

        Lamium (40)- dead nettle

        Prunella (5)- *P. vulgaris* is self-heal

        Mondarda (20)- horsemint, bee balm

        Salvia (500)- *S. officinalis* is the culinary sage

Subfamily: Ocimoideae

Hyptis (350)- desert lavender, a shrub

Ocimum (60)- *O. basilicum* is basil

Coleus (200)- *C.* x *hybridus* is coleus

Recognition Characters:

Herbs and shrubs with square stems, opposite leaves, a "minty" aroma, and 5-merous zygomorphic flowers with 2 or 4 stamens, and a deeply 4-lobed ovary with a gynobasic style. This family may be confused with the Boraginaceae which have alternate leaves and regular flowers; the Verbenaceae, which usually have a single terminal style; and the Scrophulariaceae, which often have alternate leaves, unlobed ovary, and a many - seeded fruit.

$$K^{(5)} \; Cz^{(5)} \; A^{2 \text{ or } 2+2} \; \underline{G}^{(2)}$$

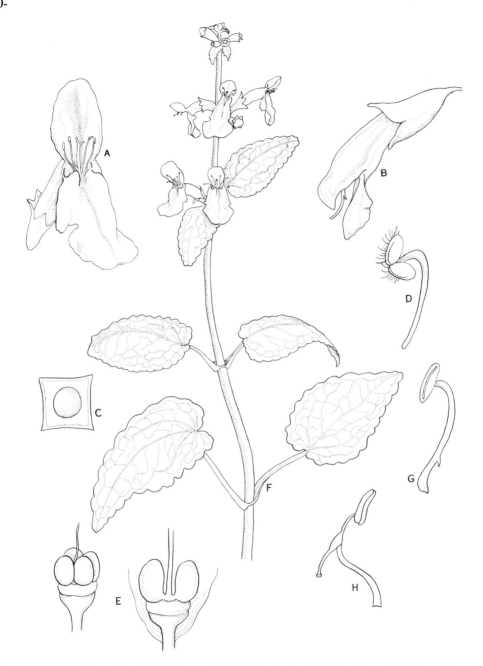

Fig. 86. **Labiatae.** a, flower showing bilabiate corolla; b, *Salvia* showing bilabiate calyx and bilabiate corolla; c, stem in x-2; d, stamen with two well-developed anther-halves; e, gynoecium showing deeply 4-lobed ovary; f, habit of a typical mint; g, stamen with one well-developed anther-half; h, stamen with one well-developed anther-half and an elongate connective arm.

PLANTAGINACEAE
Plantain Family

Annual or perennial herbs. Leaves basal, usually alternate, venation apparently parallel, bases sheathing, exstipulate. Flowers usually bisexual, actinomorphic, 4-merous, in bracteate scapose spikes or heads. Calyx of 4 united, membranous sepals, tubular. Corolla of 4 united membranous or papery petals. Androecium of 4 epipetalous stamens, anthers usually exserted on long filaments. Gynoecium of 2 united carpels, [1] 2 [4] locules, placentation usually axile, ovary superior. Fruit a circumscissile capsule or a nut surrounded by a persistent calyx. Seeds often mucilaginous.

3/270; cosmopolitan. *Plantago* is widespread in the U.S.; *Littorella* is an aquatic found in New England and Minnesota. Of little economic importance. Several species of *Plantago* are weeds and one *(P. psyllium)* is the source of seeds used to make a laxative.

Selected Genera:

Plantago (260)- plantain; a common lawn weed

Littorella (3)- N. Amer., Europe, and S. Amer.

Bougueria (1)- *B. nubicola* of Peru and Argentina

Recognition Characters:

(Excluding *Littorella*) scapose herbs with rosettes of parallel veined leaves and bracteate 4-merous flowers with membranous corollas, often exserted stamens, and mucilaginous seeds.

$$K^{(4)} \quad C^{(4)} \quad A^4 \quad \underline{G}^{(2)}$$

Fig. 87. **Plantaginaceae.** a & b, *Plantago* plant and flower. **Oleaceae.** c & d, *Fraxinus* leaves and samara.

OLEACEAE
Olive Family

Shrubs and trees. Leaves usually opposite, simple or pinnately compound, exstipulate. Flowers usually bisexual, actinomorphic. Calyx of 4 united sepals. Corolla of 4 united petals. Androecium of 2 stamens, an unusual feature in woody plant of North America. Gynoecium of 2 united carpels, the placentation axile, the ovules typically 2 per locule, the ovary superior.

29/600; cosmpolitan, particularly well represented in temperate and tropical Asia. Five genera occur in the U.S. The family is of considerable economic importance because of the olive, timber, and several ornamentals. The family is one of the few with flowers having 4 sepals and petals and 2 stamens.

Selected Genera:

Olea (20)- *O. europea* is the olive

Syringa (30)- lilac

Ligustrum (40)- privet

Chionanthus (3-4)- fringe-tree

Fraxinus (50-60)- ash; occasionally apetalous

Jasminum (200)- jasmine

Forsythia (5)- golden bells

Recognition Characters:

Trees and shrubs, typically with opposite leaves. Flowers with 4 sepals and petals, 2 stamens, and 2 united carpels.

$$K^{(4)} \; C^{(4)} \; A^2 \; \underline{G}^{(2)}$$

SCROPHULARIACEAE
Figwort or Snapdragon Family

Mostly herbs or subshrubs, rarely trees or lianas. Leaves alternate, opposite, or rarely whorled, simple, entire to pinnately dissected, exstipulate. Flowers usually bisexual, zygomorphic (some nearly actinomorphic), 5-merous, in various inflorescences. Calyx of 4 or 5 united sepals. Corolla of 4 or 5 united petals, tube long or short, often bilabiate. Androecium usually of 4 didynamous stamens, less frequently 2 or 5, with the fifth stamen often staminodial. Gynoecium of 2 united carpels, 2 locules, placentation axile, ovules numerous, ovary superior. Fruit a capsule, either loculicidal or septicidal; or a berry.

220/3000; cosmopolitan. About 40 genera are native to the U.S. The family is of economic importance because of cardiac glycosides derived from the foxglove and many fine ornamentals. As treated here, the family includes the Rhinanthaceae, Pedicullariaceae, Antirrhinaceae, Veronicaceae, Verbascaceae, and Paulowniaceae.

Selected Genera:

A. Posterior corolla teeth covering lateral ones in bud

Subfamily: Verbascoideae (5 stamens; all leaves usually alternate)

Verbascum (320)- mullein

Subfamily: Scrophularioideae (5th stamen reduced; lower leaves often opposite)

Mimulus (120)- monkey flower

Penstemon (300)- beardtongue

Scrophularia (150)- figwort, bee plant

Collinsia (20)- Chinese houses, blue lips

Calceolaria (500)- slipper flower; orchid-like

Linaria (150)- *L. vulgaris* is butter-and-eggs

Antirrhinum (42)- *A. majus* is the snapdragon

A. Posterior corolla teeth covered by one or both lateral teeth in bud

Subfamily: Rhinanthoideae

    Digitalis (21)- foxglove, the source of glycosides

    Veronica (300)- speedwell; flowers almost regular

    Hebe (70-80)- hebe, ornamental shrubs

    Castilleja (200)- Indian paintbrush

    Orthocarpus (30)- owl clover

    Pedicularis (600)- lousewort, elephant heads

Recognition Characters:

(Ours) herbs and shrubs' alternate or opposite leaves, 5-merous zygomorphic flowers with 4 stamens and sometimes a 5th staminode. The figworts may be confused with the Labiatae, which have a deeply 4-lobed ovary, gynobasic style, and 1-seeded fruits; and the Solanaceae, which are typically actinomorphic and with 5 stamens. The Scrophulariaceae are compared with closely related families in Table 16 below.

$$K^{(5)} \; Cz^{(5)} \; A^{[2]\,4\,[5]} \; \underline{G}^{(2)}$$

Table 16. A comparison of the Scrophulariaceae, Gesneriaceae, Bignoniaceae, Acanthaceae, and Lentibulariaceae.

Family	Androecium	Locules	Placentation	Miscellaneous
Scrophulariaceae	[2] 2+2 [5]	2	axillary	
Gesneriaceae	[2] 2+2 [5]	1	parietal	not native to U.S.
Bignoniaceae	[2] 2+2	2	axillary	seeds often winged
Acanthaceae	[2] 2+2 [5]	2	axillary	showy bracts
Lentibulariaceae	2	1	free-central	insectivorous

Fig. 88. **Scrophulariaceae.** a & b, *Orthocarpus,* two views of flower;   c, *Scrophularia;* d, *Mimulus,* flower; e, *Scrophularia* flower (arrow indicates staminodium); f, *Veronica;* g, *Castilleja* (arrow indicates bract); h, *Linaria* (arrow indicates spur); i & j, *Allophyton.*

GESNERIACEAE
Gesneria Family

Mostly herbs, rarely large shrubs or trees, often epiphytic. Leaves usually opposite, simple, exstipulate. Flowers bisexual, zygomorphic, 5-merous, hypogynous, solitary or in cymose inflorescences. Flowers with 5 sepals; 5 united petals, usually bilabiate; 4 epipetalous, didynamous stamens, sometimes 2 or 5; 2 united carpels, unilocular, parietal placentae, superior ovary (sometimes half- or completely inferior). Fruit a capsule, opening variously; or a berry.

140/1800; tropical and subtropical. None is native to the U.S., but several genera are encountered in cultivation.

Selected Genera:

Saintpaulia (19)- African violet

Columnea (160)- columnea; widely cultivated

Episcia (35)- carpet plant, lovejoy

Sinningia (15)- gloxinia

Recognition Characters:

(Ours) cultivated herbs with showy, bilabiate, scroph-like flowers. See also Table 16.

$$K^{5} \quad Cz^{(5)} \quad A^{2+2} \quad \underline{G}^{(2)}$$

OROBANCHACEAE
Broomrape Family

Annual or perennial herbaceous root parasites, commonly lacking chlorophyll. Leaves alternate, scale-like. Flowers bisexual, zygomorphic, 5-merous, hypogynous, in terminal racemes or spikes. Flowers typically with 2-5 united sepals, 5 united petals, 4 didynamous stamens, and 2 united carpels. Fruit a loculicidal capsule.

13/180; north temperate. The four genera listed below are native to the U.S. The family is of no direct economic importance.

Selected Genera:

Orobanche (100)- broom rape

Boschniakia (2)- ground cone

Epifagus (1)- *E. virginiana* is beech drops, a beech parasite

Conopholis (2)- squaw root

Recognition Characters:

Red, brown, purple or whitish herbaceous root parasites with scroph-like flowers. See also Table 16.

$$K^{(2-5)} \quad Cz^{(5)} \quad A^{2+2} \quad \underline{G}^{(2)}$$

BIGNONIACEAE
Bignonia Family

Mostly trees, shrubs, and vines, rarely herbs. Leaves usually opposite, simple or compound, exstipulate. Flowers bisexual, zygomorphic, 5-merous, in cymes. Calyx of 5 united sepals. Corolla of 5 united petals. Androecium of 4 didynamous, epipetalous stamens, less frequently 2. Gynoecium of 2 united carpels, 2-locular, placentation axile, ovary superior. Fruit a loculicidal or septicidal capsule, often elongate. Seeds usually winged, sometimes comose.

120/800; primarily tropical. Seven genera are native to the U.S., mostly in the southern half of the country. The family is the source of timber and several ornamentals.

Selected Genera:

Crescentia (5)- calabash tree

Kigelia (10)- sausage tree; bat pollinated

Tabebuia (100)- West Indian boxwood; roble blanco

Spathodea (2)- South African tulip tree

Jacaranda (40)- jacaranda, a popular ornamental

Catalpa (13)- Indian bean, catalpa

Campsis (5)- trumpet creeper

Chilopsis (1)- *C. linearis* is the desert willow

Recognition Characters:

Trees, shrubs, and woody vines with scroph-like flowers, the seeds usually winged and lacking endosperm. See also Table 16.

$$K^{(5)} \ Cz^{(5)} \ A^{2+2} \ \underline{G}^{(2)}$$

ACANTHACEAE
Acanthus Family

Perennial herbs, shrubs, rarely trees; some spiny. Leaves opposite, simple, exstipulate. Cystoliths (deposits of calcium carbonate) are often present as streaks or protuberances in the vegetative parts of the plant. Flowers bisexual, zygomorphic, often subtended by showy bracts, in cymose or racemose inflorescences. Calyx with 4 or 5 united sepals. Corolla of 4 or 5 united petals, commonly bilabiate. Androecium typically of 4 didynamous stamens, epipetalous, less frequently 2 or 5. Gynoecium of 2 united carpels, bilocular, placentation axile, ovary superior. Fruit usually a loculicidal capsule.

250/2600; pantropical. About 15 genera occur in the U.S., mostly in the southeast and southwest. Several genera are ornamentals.

Selected Genera:

Acanthus (30)- bear's breech

Thunbergia (150)- clock vine

Aphelandra (150)- widely planted ornamental

Justicia (300)- water willow

Beloperone (60)- shrimp plant

Recognition Characters:

(Ours) herbs or shrubs with opposite leaves, cystoliths, and bracteate inflorescences of 4- or 5-merous bilabiate flowers. See also Table 16.

$$K^{(4-5)} \ Cz^{(4-5)} \ A^{2+2} \ \underline{G}^{(2)}$$

LENTIBULARIACEAE
Bladderwort Family

Annual or perennial herbs, usually insectivorous, of moist and aquatic habitats. Leaves alternate or in rosettes; in aquatic species the air leaves and water leaves usually quite different, the latter often very finely divided. Flowers bisexual, zygomorphic, in racemes, spikes, or solitary. Flowers with 2-5 sepals; 5 united petals, often bilabiate; 2 stamens, and a unilocular, bicarpellate gynoecium with free central placentation. Fruit a capsule, opening variously.

5/300; cosmopolitan. *Utricularia* and *Pinguicula* are native to the U.S. The insectivorous species, trapping by means of sticky leaves or bladders, are sometimes cultivated as oddities.

Selected Genera:

Pinguicula (35)- butterwort

Utricularia (250)- bladderwort

Genlisea (15)- a terrestrial form with bladders

Recognition Characters:

Plants of this family resemble those of the Scrophulariaceae, but may be distinguished by their insectivorous habit, reduced stamen number, and free central placentation. See also Table 16.

$$K^{2-5} \ Cz^{(5)} \ A^{2} \ \underline{G}^{(2)}$$

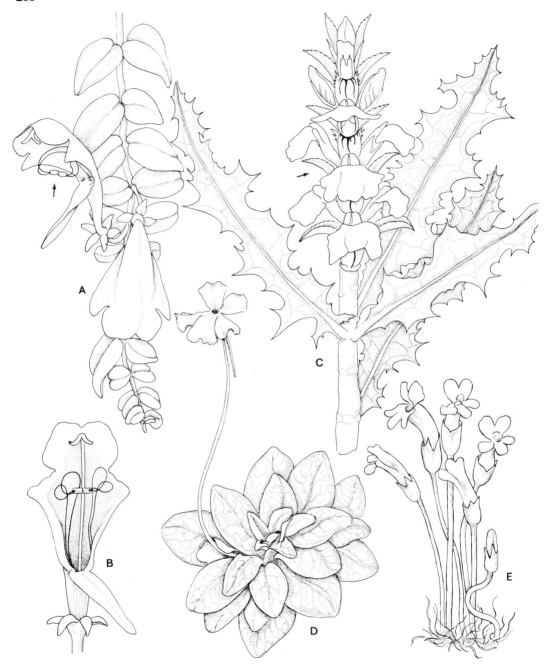

Fig. 89. **Gesneriaceae.** a & b, *Columnea* flowering branch and flower showing syngenesious stamens (indicated by arrow). **Acanthaceae.** c, *Acanthus* (arrow indicates bract subtending flower). **Lentibulariaceae.** d. *Pinguicula.* **Orobanchaceae.** e, *Orobanche.*

CAMPANULACEAE
Harebell Family

Mostly perennial herbs, sometimes shrubs, rarely trees. Sometimes with milky latex. Leaves alternate, simple, exstipulate. Flowers usually bisexual, zygomorphic or actinomorphic, 5-merous, epigynous. Calyx of [3] 5 [10] sepals. Corolla of 5 united petals. Androecium usually of 5 epipetalous stamens, sometimes united by anthers (Subfamily Lobelioideae). Gynoecium of 2, 3 or 5 united carpels, as many locules, placentation axile, ovules, numerous, ovary inferior or half - inferior. Fruit a capsule, dehiscing variously; or a berry.

70/2000; tropical, subtropical, and temperate. Twelve genera are native to the U.S. Of little economic importance; several are cultivated as ornamentals. As treated here, the family includes the Lobeliaceae.

Selected Genera:

Subfamily: Campanuloideae (actinomorphic; anthers free)

Campanula (ca. 300)- harebell, bluebell

Platycodon (1)- *P. grandiflorum* is the balloon flower

Triodanis (7)- Venus' looking glass

Subfamily: Lobelioideae (zygomorphic; anthers united)

Lobelia (380)- Indian tobacco, cardinal flower

Centropogon (226)- tropical American

Recognition Characters:

Herbs or shrubs with 5-merous, zygomorphic or actinomorphic, epigynous flowers with axile placentation and numerous ovules.

$$K^5 \; C^{(5)} \text{ or } Cz^{(5)} \; A^5 \; \overline{G}^{(2\text{-}3\text{-}5)}$$

Fig. 90. **Campanulaceae.** a & b, *Campanula* plant and flower in longitudinal section; c & d, *Lobelia* plant and flower in longitudinal section.

RUBIACEAE
Madder Family

Trees, shrubs, and herbs. Leaves opposite or whorled, simple, usually entire; stipules present and often leaf-like. Flowers bisexual, usually actinomorphic, in cymes. Calyx of 4 or 5 sepals. Corolla of 4 or 5 united petals, usually regular, but sometimes bilabiate. Androecium of 4 or 5 epipetalous stamens. Gynoecium of 2 united carpels (rarely 1-many), usually 2-locular, placentation axile or appearing basal, ovary inferior (rarely superior). Fruit a loculicidal or septicidal capsule or a berry.

500/6000-7000; pantropical. About 20 genera are native to the U.S., but most of them are herbaceous and do not give a proper impression of the family as a whole. Of economic importance because of coffee, quinine, and many ornamentals.

Selected Genera:

Hedyotis (including *Oldenlandia*) (300+)- bluets

Houstonia (35)- bluets

Pentas (32)- star-cluster, a popular ornamental shrub

Cinchona (ca. 15)- several species yield the antimalarial quinines

Cephalanthus (10)- button bush, a shrub of wet sites

Gardenia (100+)- *G. jasminoides* is the common gardenia

Coffea (50-60)- *C. arabica* and other species yield coffee beans

Galium (ca. 300)- bedstraw, one of our most common representatives

Rubia (60)- *R. tinctorum* is the source of madder, a dye

Recognition Characters:

(Ours) stipulate herbs and shrubs with opposite or whorled leaves, 4 or 5-merous flowers with an inferior ovary. The family is most easily confused with the Caprifoliaceae which usually lack stipules.

$$K^{\,4\text{-}5} \; C^{\,(4\text{-}5)} \; A^{\,4\text{-}5} \; \overline{G}^{\,(2)}$$

Fig. 91. **Rubiaceae.** a, *Galium;* b, *Cephalanthus.*

CAPRIFOLIACEAE
Honeysuckle Family

Mostly shrubs, sometimes lianas or herbs. Leaves opposite, usually simple and exstipulate. Flowers bisexual, actinomorphic or zygomorphic, in cymose inflorescences. Calyx of 5 small sepals. Corolla of 5 united petals, rotate, salverform or bilabiate. Androecium of 5 [4] epipetalous stamens. Gynoecium of 2-3-5-8 united carpels, 1-5 locules, placentation axile, ovary inferior. Fruit usually a berry or drupe, sometimes a capsule.

15/400; north temperate. Seven genera are native to the U.S. Most of the genera are in cultivation as ornamentals. As treated here, the family includes the Sambucaceae.

Selected Genera:

Lonicera (ca. 180)- honeysuckle

Diervilla (3)- bush honeysuckle

Symphoricarpos (15)- snowberry, Indian currant, coral-berry

Kolkwitzia (1)- *K. amabilis* is the beautybush

Linnaea (1)- *L. borealis* is the twinflower

Viburnum (ca. 120)- snowball tree, nannyberry, wayfaring-tree

Sambucus (25)- elderberry; fruits made into a wine

Triosteum (8)- horse gentian

Recognition Characters:

Exstipulate shrubs with opposite leaves, 5 or 4-merous, epigynous flowers, and a multicarpellate gynoecium. The family is most easily confused with the Rubiaceae. The two families are separated by the presence or absence of stipules.

$$K^5 C^{(5)} \text{or } Cz^{(5)} \quad A^5 \quad \overline{G}^{(2-8)}$$

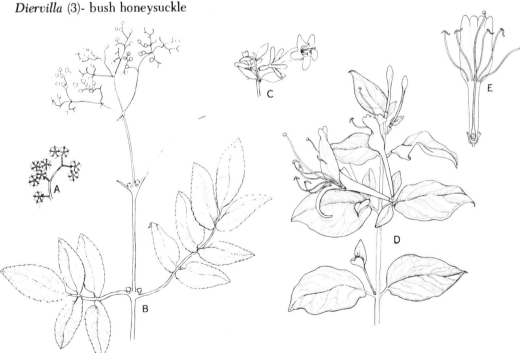

Fig. 92. **Caprifoliaceae.** a & b, *Sambucus;* c, *Abelia;* d & e, *Lonicera.*

VALERIANACEAE
Valerian Family

Annual or perennial herbs. Leaves opposite or in basal rosettes, often pinnately divided, bases often sheathing, exstipulate. Flowers bisexual or unisexual (the species dioecious), zygomorphic, in cymose panicles. Calyx rudimentary. Corolla of 5 united petals, often spurred. Androecium of 1-4 epipetalous stamens. Gynoecium of 3 united carpels, one functional. Fruit an achene.

13/400; north temperate and Andean. The first three genera listed below are native to the U.S. The family is of no economic importance; a few are ornamental.

Selected Genera:

Plectritis (15)- mostly in the West

Valeriana (250)- valerian root, a sedative

Valerianella (60)- corn salad

Centranthus (10)- *C. ruber* is the red valerian

Nardostachys (3)- *N. jatamansii* is the spikenard

Recognition Characters:

Herbs, the leaves opposite or in basal rosettes. Flowers zygomorphic, spurred or saccate; androecium of 1, 2 or 4 stamens; gynoecium tricarpellate and inferior.

$$K^0 \ Cz^{(5)} \ A^{1-4} \ \overline{G}^{(3)}$$

DIPSACACEAE
Teasel Family

Mostly herbaceous. Leaves opposite, exstipulate. Flowers bisexual, zygomorphic, epigynous, in dense cymose heads, each flower subtended by 2 fused bracteoles, thus forming an epicalyx. Calyx 5-merous, cuplike, tubular or divided into several segments. Corolla of 4 or 5 united petals. Androecium of 4 epipetalous stamens. Gynoecium of 2 united carpels, unilocular, ovary inferior. Fruit an achene.

10/270; Old World. None is native to the U.S., although *Dipsacus* is widely naturalized and has become quite weedy. The dried inflorescence is often sprayed some frightful color and put into floral arrangements.

Selected Genera:

Dipsacus (15-20)- teasel, fuller's teasel; used to raise nap

Knautia (40)- blue buttons

Scabiosa (ca. 80)- pincushion-flower, mourning bride

Recognition Characters:

Herbs with opposite, exstipulate leaves. Flowers 5-merous, epigynous, in dense involucrate heads. *Dipsacus*, the most commonly encountered member of the family, superficially resembles a thistle or *Eryngium* of the Umbelliferae.

$$K^{(5)} Cz^{(5)} \ A^4 \ \overline{G}^{(2)}$$

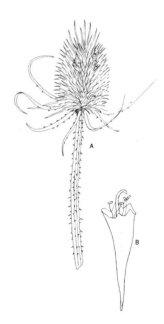

Fig. 93. **Dipsacaceae.** a, inflorescence of *Dipsacus*; b, flower detail.

COMPOSITAE (= ASTERACEAE)
Aster Family

Mostly Herbaceous, less frequently shrubs, lianas, and trees. Sap watery (Subfamily Tubuliflorae) or milky (Subfamily Liguliflorae). Leaves usually alternate, frequently in basal rosettes, simple or compound, exstipulate. Flowers typically in a head or capitulum, the heads themselves further organized into racemes, corymbs, panicles, or even compound capitula.

A head is composed of many tiny flowers or **florets** inserted on a conical, hemispheric or flattened **common receptacle** or **disk**. The entire capitulum is surrounded by an involucre of bracts called **phyllaries**. These may occur in 1-several series or they may be imbricate. A second type of bract (**chaff**) may also occur. This structure is found subtending individual flowers on the common receptacle. If the bracts are present, the receptacle is **chaffy**; if absent, it is **naked**. The receptacle may also bear hairs, bristles, or be pitted.

The small flowers are bisexual or unisexual (the species monoecious or dioecious), actinomorphic or zygomorphic, 5-merous, and epigynous. Only one perianth series is well-developed. The calyx, referred to in this family as the **pappus**, is rudimentary or absent. When present, it takes the form of a low crown or a series of capillary bristles, plumose bristles, stout awns or scales (paleae). The corolla is 5-merous and sympetalous. It may be actinomorphic in those flowers referred to as **tubular** or zygomorphic in those called **ligulate** or **bilabiate**. The androecium is composed of 5 stamens, these united by their anthers (sygenesious) to form a tube around the style. The gynoecium is bicarpellate, unilocular, the ovary inferior. Fruit an achene or cypsela.

Although the above generalized floral description is useful, it does obscure the fact it is possible to recognize four flower types.

1). Tubular or Disk flower
KxC(5)A5G(2)or G0
Note that the corolla is actinomorphic and that the flower may be perfect or functionally staminate. Tubular flowers may be the only type found in a capitulum or they may occupy the center of the inflorescence and be surrounded by flowers of the next type.

2). Ligulate or Ray flower (Type A)
KxCz(5)A0G(2)or G0
Here the corolla is zygomorphic and the flowers are either functionally pistillate or neuter. The corolla is produced into an elongate structure called a **strap** or **ligule**. Its apex is typically rounded. Such flowers occupy the periphery of the head, surrounding the central tubular flowers. Note how the two types may sexually complement one another.

3). Ligulate or Ray flower (Type B)
KxCz(5)A5G(2)
In this second type of ligulate flower, the elongate corolla is typically 5-toothed at its apex and the flower is perfect. When present, it will be the only type to be present on the receptacle.

4). Bilabiate flower
KxCz(5)A5G(2)
This unusual flower type with a bilabiate corolla is restricted to the tribe mutisieae, which are represented in the U.S. mostly by ornamentals.

Ignoring the rare bilabiate flower, three types of capitula may be recognized based upon the flowers they contain. A head may contain only tubular flowers, or central tubular flowers and peripheral ligulate flowers (Type A), or only ligulate flowers (Type B). The first and second capitulum types are found in plants of the subfamily Tubuliflorae, while the third condition is

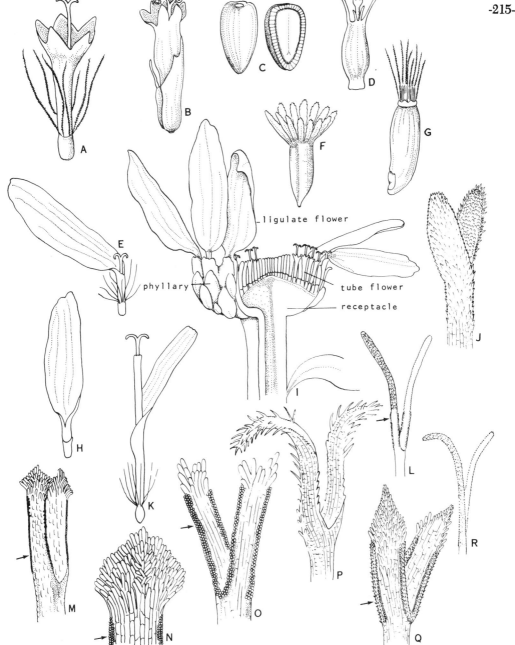

Fig. 94. **Compositae.** a, tubular flower with pappus of plumose bristles; b, tubular flower with chaff; c, generalized achene, exterior and longitudinal section; d, achene with pappus of awns; e, pistillate ligulate flower with pappus of capillary hairs; f, achene with pappus of paleae; g, achene with low crown and plumose bristles; h, sterile ligulate flower; i, generalized inflorescence; j, style, Anthemideae; k, perfect ligulate flower with pappus of capillary hairs; l, style, Eupatorieae; m, style, Senecioneae; n, style detail, Senecioneae; o, style, Inuleae; p, style, Heliantheae; q, style, Astereae; r, style, Vernonieae. (Arrows indicate stigmatic surfaces; drawings j; l-r after Wood, 1974).

Table 17. Comparison of the tribes of the Compositae.

	Latex	Aroma[1]	Flowers[2]	Anthers	Stigma-Style	Pappus	Receptacle
Subfamily: Tubuliflorae							
Tribe:							
Heliantheae	-	-	T, T+La	blunt or acute	crown of hairs below stigma	various or none	chaffy or naked
Astereae	-	-	T, T+La	blunt	stigma flat, hairy terminal	mostly awns or bristles	naked
Anthemideae	-	+	T, T+La	blunt or acute	stigmas flattened	usually none	naked or chaffy
Senecioneae	-	-	T, T+La	blunt or acute	branches flat and truncate	capillary bristles	usually naked
Inuleae	-	-	T	tailed	branches obtuse & truncate	capillary or none	naked or chaffy
Calenduleae	-	-	T+La	pointed	branches bifid or entire	usually none	naked
Arctotoideae	-	-	T+La	pointed	branches connate to tip	paleaceous or lacking	naked or chaffy
Eupatorieae	-	-	T	blunt	elongate, thickened above	usually setose	usually naked
Vernonieae	-	-	T	tailed-pointed	stigmas long-pointed	usually setose	usually naked
Cynareae	-	-	T	tailed	circle of hairs below stigma	bristly or plumose bristles	many bristles
Mutisieae	-	-	B, B+La	tailed	branches short and blunt	bristles	usually naked
Subfamily: Liguliflorae							
Tribe:							
Cichorieae	+	-	Lb	sagittate	narrow and cylindric	bristles	chaffy or naked

1) Many members of the family are somewhat aromatic; those of the Anthemideae are especially so.

2) T= tubular; La= ligulate flower, Type A; Lb= ligulate flower, Type B; B= bilabiate

restricted to the Liguliflorae. See Table 17 for a comparison of the tribes of these two subfamilies.

The Compositae, with 920 genera and 19,000 species, are second only to the Orchidaceae in size. Composites are found in every part of the world, but are infrequent in the tropical rain forest. Aquatic, marsh, and epiphytic species are also uncommon. The vast majority are herbaceous. Over 200 genera occur in the U.S. The family is of little economic importance, certainly considering its size. Many species are ornamental, a few (lettuce, sunflower fruits, artichoke, etc.) are edible, *Carthamnus* yields the safflower oil, and one (*Brachylaena*) is a timber tree.

Selected Genera:

Subfamily: Tubuliflorae (at least some tubular flowers present; sap watery)

Tribe: Heliantheae (including Helenieae)

Only tubular or tubular and ligulate flowers; style with crown of hairs below stigma; anthers rounded at base; pappus not hairy; receptacle naked or chaffy; leaves often opposite

Helianthus (100)- sunflower

Cosmos (29)- cosmos

Rudbeckia (30)- coneflower, black-eyed susan

Ratibida (6)- coneflower

Echinacea (5)- purple coneflower

Balsamorhiza (12)- balsam root

Silphium (20)- rosinweed, compass plant

Parthenium (16)- guayule

Iva (15)- poverty weed

Ambrosia (including *Franseria*) (ca. 50)- ragweed

Xanthium (5)- cocklebur

Zinnia (17)- zinnia

Helenium (40)- sneezeweed

Gaillardia (35)- blanket flower

Tagetes (35)- marigold

Tribe: Astereae

All or only disc flowers tubular; anthers with blunt bases; stigmas flattened with marginal rows of papillae and terminal hairy segment

Aster (ca. 500)- aster

Grindelia (50) tarweed

Solidago (ca. 100)- goldenrod

Chrysothamnus (12)- rabbit brush

Bellis (10)- English daisy

Erigeron (ca. 250)- flea bane

Conyza (ca. 100)- mares tail

Baccharis (ca. 400)- groundsel-bush

Tribe: Anthemideae

Flowers all tubular or tubular and ligulate; phyllaries with membranous edges at tips; pappus lacking or rudimentary; often aromatic plants

Anthemis (110)- mayweed, dog fennel, stinkweed

Achillea (100+)- yarrow

Matricaria (50)- pineapple weed

Chrysanthemum (ca. 200)- chrysanthemum

Artemisia (ca. 250)- sagebrush, wormwood

Tribe: Inuleae

Tubular flowers only, except *Inula*; anthers tailed

Antennaria (ca. 50)- pussy toes, ladies-tobacco

Anaphalis (50)- pearly everlasting

Gnaphalium (ca. 150)- cudweed

Leontopodium (50)- edelweiss

Tribe: Senecioneae

As in Heliantheae, but pappus hairy

Petasites (20)- colts foot

Cineraria (40)- cultivated cineraria

Senecio (includ. *Kleinia*) (1,500+)- groundsel

Tribe: Calenduleae

Usually pistillate ray flowers and staminate disc flowers; anthers pointed; receptacle naked; no pappus; Old World only

Dimorphotheca (7)- cape marigold

Calendula (15)- pot marigold

Tribe: Arctotoideae

Style thickened or with a ring of hairs below stigma; involucre and leaves often spiny; mostly South African

Gazania (20)- gazania

Tribe: Eupatorieae

Tubular flowers only; never true yellow; anthers blunt; stigmas long, but with blunt or flattened tips; leaves mostly opposite

Ageratum (35)- ageratum

Eupatorium (600)- thoroughwort, boneset, joe-pye weed

Liatris (32)- blazing star, gayflower

Tribe: Vernonieae

Tubular flowers only; never yellow; anthers sagittate, pointed or rarely caudate; stigmas long-pointed

Vernonia (ca. 600)- ironweed

Tribe: Cynareae

Styles thickened or with a circle of hairs below stigma; tubular flowers only; anthers usually tailed; receptacle with spiny involucre

Arctium (6)- burdock

Carduus (100)- musk thistle

Cirsium (250)- thistle

Cynara (10)- artichoke

Silybum (2)- milk thistle

Centaurea (500)- star thistle, knapweed, bachelors-button

Echinops (ca. 120)- globe thistle

Tribe: Mutisieae

Tubular or tubular and ray flowers; both usually bilabiate

Gerbera (50)- transvaal daisy

Subfamily: Liguliflorae (all flowers ligulate; sap milky)

Tribe: Cichorieae

Cichorium (8)- chicory, used in coffee

Tragopogon (45)- goats beard, salsify

Sonchus (70)- sowthistle

Hieracium (800)- hawkweed

Lactuca (100)- lettuce

Taraxacum (70)- dandelion

Crepis (200)- hawksbeard

Recognition Characters:

Herbs and shrubs with involucrate heads of small, 5-merous, sympetalous flowers. The calyx is represented by a pappus, a series of bristles, hairs or scales. Stamens 5, united by their anthers. Gynoecium bicarpellate, inferior, producing a single achene at maturity.

$$K^x \, C^{(5)} \text{ or } Cz^{(5)} \, \overset{\frown}{A^5} \, \overline{G}^{(2)}$$

Liliopsida

Order: Family	Common Name	Genera/Species	Distribution	Form	K
Alismatales					
Butomaceae	Flowering Rush	1/1	Euras.	H	3
Alismataceae	Arrowhead	13/90	Cosmo.	H	3
Hydrocharitales					
Hydrocharitaceae	Frog's-bit	15/100	Trop. & temp.	H	3
Najadales					
Scheuchzeriaceae	Scheuchzeria	1/2	N. temp.	H	3
Juncaginaceae	Arrow grass	4/26	N. & S. Temp.	H	3
Najadaceae	Water nymph	1/50	Cosmo.	H	0?
Potamogetonaceae	Pond weed	2/100	Cosmo.	H	4?
Ruppiaceae	Ditch grass	1/2	Temp. & trop.	H	0
Zannichelliaceae	Grass wrack	3/6	Cosmo.	H	3 or 0
Zosteraceae	Eel grass	3/18	Marine waters	H	0

Alismatidae

C	A	G	Fruit Type(s)	Miscellaneous Comments
3	9	6	fol	Naturalized in NE & NC U.S.
3	6-∞	<u>6-∞</u>	ach	Marsh or aquatic plants
3	1-∞	(2-x)	ber-like	Aquatics of freshwater or marine sites
3	6	<u>6 or 3</u>	schizo or fol	Slender herbs of *Sphagnum* bogs
3	6-4,1	<u>3 or 6</u>	fol, nut	Including the Lilaeaceae
0	1	<u>1</u>	ach	Submersed annuals
0	4	<u>4</u>	drplt	Submersed perennials
0	2	<u>4</u>	nutl	Differ from pond weeds in stamen number
0	1-3	<u>1-9</u>	nutl?	Submersed perennials
0	1	<u>1</u>	ach	*Zostera* and *Phyllospadix* off U.S. coasts

The Alismatidae consist of four orders, fourteen families, and fewer than 500 species. It is the smallest subclass of the Liliopsida. Most are aquatic or semiaquatic herbs whose flowers have many stamens and an apocarpous gynoecium. Cronquist considers the group a near - basal side branch which has retained many primitive features.

In the Alismatales the perianth is usually well - developed, biseriate, and 3 - parted. The Hydrocharitales are peculiar in being epigynous. In the Najadales the perianth is not clearly distinguished into sepals and petals.

BUTOMACEAE
Flowering Rush Family

Aquatic herbs. Leaves erect, linear, not differentiated into lamina and petiole. Flowers bisexual, actinomorphic, 3-merous, in a scapose umbel. Flowers with 6 tepals, 9 stamens, and 6 separate or basally connate carpels; placentation parietal, the ovary superior. Fruit a follicle.

1/1; Eurasian. Although not native to the U.S., *Butomus* has become naturalized in several areas of the Northeast. Other genera traditionally placed in the Butomaceae are now transferred to the Limnocharitaceae.

Genus:

Butomus (1)- *B. umbellatus* is the flowering rush

Recognition Characters:

Aquatic herbs with erect linear leaves and lily-like flowers with 6 petaloid tepals, 9 stamens, and 6 carpels.

$$K^3 C^3 A^9 \underline{G}^6$$

ALISMATACEAE
Arrowhead Family

Annual or perennial marsh and aquatic herbs. Leaves basal, erect or floating, petiole long, bases sheathing, blades linear to sagittate or hastate. Flowers bisexual or unisexual (the species monoecious or dioecious). Actinomorphic, bracteate, hypogynous, in racemes or panicles. Calyx of 3 separate green sepals. Corolla of 3 separate petals, often white. Androecium of [3] 6-many stamens. Gynoecium of 6-many separate carpels (spirally arranged in primitive genera), unilocular, one ovule per locule, ovary superior. Fruit usally an achene.

13/90; cosmopolitan. Five genera are native the the U.S. Of little economic importance; *Sagittaria* rhizomes are eaten; some species are cultivated as ornamentals. The spelling of the family name is conserved over the Alismaceae.

Selected Genera:

Alisma (10)- water plaintain

Echinodorus (25-30)- burhead

Sagittaria (20)- arrowhead

Recognition Characters:

Marsh and aquatic herbs, sometimes with arrowhead-shaped leaves, with flowers having 3 green sepals, 3 white petals, and an apocarpous gynoecium, usually with many carpels. The family may be easily confused with the Ranunculaceae. One intermediate genus, *Ranalisma,* is considered in this family rather than the Ranunculaceae because it has one cotyledon and no endosperm. The flowers are essentially identical in the two families.

$$K^3 C^3 A^{6-\infty} \underline{G}^{6-\infty}$$

HYDROCHARITACEAE
Frog's-bit Family

Aquatic herbs, both freshwater and marine, with submersed ribbon-like leaves (except *Elodea*). Flowers unisexual (the species dioecious) or bisexual, actinomorphic, umbellate or solitary, subtended by spathaceous bracts. Calyx of 3 sepals, these often green. Corolla of 3 petals. Androecium of 1-many stamens. Gynoecium of 2-x united carpels, ovary inferior. Fruit berry-like, irregularly dehiscent.

15/100; tropical and temperate. Five genera are native to the U.S., 3 of them (*Hydrilla*, *Halophila*, and *Thalassia*) in the marine waters off Florida. Several species are cultivated as aquarium plants.

Selected Genera:

Elodea (including *Anacharis*) (15)- waterweed

Vallisneria (6-10)- tape grass, eel grass

Thalassia (1)- *T. testudinum* is turtle grass, a submersed marine plant off the south Florida coast

Recognition Characters:

Freshwater and marine submersed aquatics with ribbon-like leaves (broader in *Elodea*). Flowers subtended by spathaceous bracts. Flowers like those of the Butomaceae, but syncarpous and epigynous.

$$K \; ^3 \; C \; ^3 \; A \; ^{1-\infty} \; \overline{G} \; ^{(2-x)}$$

JUNCAGINACEAE
Arrow Grass Family

Perennial herbs of freshwater and salt marshes. Leaves linear, sheathing. Flowers bisexual or unisexual (the species dioecious or polygamous), small, actinomorphic, hypognous, in racemes or spikes. Perianth of 6 tepals, 3 + 3. Androecium of 6 [4] stamens. Gynoecium of 3-6 separate or basally connate carpels, separating at maturity. Fruit a follicle. In *Lilaea* the unisexual flowers are greatly reduced, the staminate flower consisting of a single stamen, the pistillate flower of a supposedly 3-carpellate gynoecium.

4/26; tropical, temperate, and subarctic regions. *Triglochin* and *Lilaea* occur in the U.S. Of no economic importance. As treated here, the family includes the Lilaeaceae.

Selected Genera:

Triglochin (15)- arrow grass; toxic because it produces HCN

Lilaea (1)- *L. scilloides* is the flowering quillwort

Recognition Characters:

Perennial herbs of freshwater or brackish sites. The leaves linear and basal. The flowers small (naked in *Lilaea*), 3-merous, in racemes or spikes; fruit a follicle.

$$K \; ^3 \; C \; ^3 \; A \; ^{6[4]} \; \underline{G} \; ^{3-6}$$

NAJADACEAE
Water Nymph Family

Submerged aquatic annuals. Leaves linear, subopposite or verticillate, toothed, sheathing at the base. Flowers unisexual, the species usually monoecious. Staminate flowers composed of a single stamen, enclosed by a tiny bract, possibly representing a perianth. Pistillate flowers of one carpel, naked or enclosed by a membranous bract. Pollination occurs underwater. Fruit an achene.

1/50; cosmopolitan. Several species are native to the U.S. Of no economic importance.

Genus:

Najas (50)- water nymph

Recognition Characters:

Submersed annual aquatics. Linear leaves. Flowers unisexual, reduced to a single stamen or carpel, these enclosed by a tiny membranous bract.

$$K \; ^0 \; C \; ^0 \; A \; ^1 \; G \; ^0 \quad \text{(staminate flower)}$$

$$K \; ^0 \; C \; ^0 \; A \; ^0 \; \underline{G} \; ^1 \quad \text{(pistillate flower)}$$

POTAMOGETONACEAE
Pond Weed Family

Perennial aquatic herbs of freshwater sites, rarely of bogs and marshes. Stems often jointed. Leaves opposite or alternate, 2-ranked, blades floating or submersed, basally sheathing. Flowers bisexual or unisexual (the species usually monoecious), small, actinomorphic, in axillary spikes borne on well-developed bracteate peduncles. Perianth of 4 distinct, clawed segments. Androecium of 4 stamens, inserted on the perianth segments. Gynoecium of 4 separate carpels. Fruit of 1-4 drupelets or achenes.

2/100; cosmopolitan. *Potamogeton* is native to the U.S. Of no direct economic importance.

Selected Genus:

Potamogeton (100)- pondweed

Recognition Characters:

Perennial aquatics of freshwater sites. Leaf blades floating or submersed. Flowers 4-merous.

$$K^4 \; C^0 \; A^4 \; \underline{G}^4$$

Fig. 95. **Juncaginaceae.** a-d, *Triglochin* plant, flower with two series of perianth-like appendages, flower with maturing anthers, and fruit. **Alismataceae.** e, *Sagittaria*. **Butomaceae.** f, *Butomus.* **Hydrocharitaceae.** g, *Elodea.* (Drawings b & c after Mason, 1957).

Liliopsida

Order: Family	Common Name	Genera/Species	Distribution	Form	K
Commelinales					
Xyridaceae	Yellow-eyed grass	2/250	Trop. & subtrop.	H	3
Mayacaceae	Bog moss	1/10	Trop. Am. & Afr.	H	3
Commelinaceae	Spiderwort	40/600	Trop. & subtrop.	H	3
Eriocaulales					
Eriocaulaceae	Pipewort	13/1150	Trop. & subtrop.	H	2-3
Juncales					
Juncaceae	Rush	9/400	Cool temp; subarc.	H	3
Cyperales					
Cyperaceae	Sedge	90/4000	Cool temp; subarc.	H	x
Gramineae	Grass	600/10,000	Cosmo.	H	2-3
Typhales					
Sparganiaceae	Bur-reed	1/20	Chiefly N. temp.	H	3-6
Typhaceae	Cattail	1/15	Temp. & trop.	H	x
Bromeliales					
Bromeliaceae	Pineapple	50/2000	Chiefly NW trop.	H	3
Zingiberales					
Strelitziaceae	Bird of Paradise	3/7	OW & NW trop.	H	3
Musaceae	Banana	2/70	Afr., As., Austr.	H	3
Zingiberaceae	Ginger	49/1500	Trop. & subtrop.	H	3
Marantaceae	Arrowroot	30/400	Chiefly Am. trop.	H	3

Commelinidae

C	A	G	Fruit Type(s)	Miscellaneous Comments
3	3	(3)	cap	*Xyris* from N. Jersey southward
3	3	(3)	cap	Atlantic and Gulf coastal plains
3	6	(3)	cap	Filaments often with conspicuous hairs
2-3	4 or 6	(2-3)	cap	Not found in the western U.S.
3	6	(3)	cap	Small grass-like herbs
0	3	(2-3)	ach, nutl	Grass-like; stems often 3-sided
0	3	(2-3)	cary	Stunningly beautiful plants
0	3-6	1	nut- or drp-like	Unisexual aquatics
0	2-5	1	ach, nutl	Large marsh plants with spikes of flowers
3	6	(3)	ber, cap, mult	Inflorescences usually quite conspicuous
2+1z	5	(3)	cap	None is native, but *Strelitzia* cultivated
2+1z	5+1	(3)	ber	None is native, but *Musa* is cultivated
2+1	1+1-2	(3)	cap	None is native, but several in cultivation
3	1+1-2	(3)	cap	*Thalia* occurs in the southeastern U.S.

Eight orders, 25 families, and almost 19,000 species comprise the Commelinidae. Most are terrestrial herbs with a syncarpous gynoecium. Some members of the subclass have a well - developed perianth, clearly differentiated into a calyx and corolla. In others, the perianth is greatly reduced.

The Commelinales have conspicuous flowers with a perianth differentiated into sepals and petals. In the Eriocaulales the perianth is reduced or absent. The Juncales are characterized by six sepaloid tepals, 3 - carpellate gynoecium, capsules, and anemophily. In the Cyperales, the flowers are reduced, wind pollinated, and usually occur in spikelets. The Typhales are aquatic plants with reduced, unisexual, wind pollinated flowers. The Bromeliales are insect pollinated terrestrial or epiphytic herbs with a superior or inferior ovary. The Zingiberales are pinnately - veined herbs whose flowers are zygomorphic, epigynous, and have 1-5 functional stamens.

COMMELINACEAE
Spiderwort Family

Perennial or annual herbs with succulent stems and ± swollen nodes. Often mucilaginous. Leaves alternate, entire, with a closed basal sheath. Flowers bisexual, actinomorphic (rarely zygomorphic), borne in boat-shaped bracts. Calyx of 3 green sepals. Corolla of 3 commonly blue petals. Androecium of 6 stamens, the filaments often with brightly-colored hairs. Gynoecium of 3 united carpels; 3-locular, placentation axile, ovary superior. Fruit a loculicidal capsule.

40/600; primarily tropical and subtropical. Five genera are native to the U.S. Several are in cultivation as ornamentals.

Selected Genera:

Commelina (150)- day-flower

Tradescantia (30-50)- spiderwort

Setcreasea (8)- purple heart

Zebrina (1-2)- wandering jew

Rhoeo (1)- *R. spathacea* is Moses-in-the cradle

Recognition Characters:

Herbs with succulent, mucilaginous stems, prominent closed sheaths, 3 green sepals, 3 often blue petals, and 6 stamens with hairy filaments.

$$K \ 3 \ C \ 3 \ A \ 6 \ \underline{G} \ ^{(3)}$$

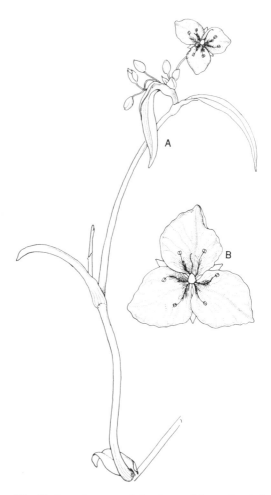

Fig. 96. **Commelinaceae.** a & b, plant and flower detail of *Tradescantia.*

JUNCACEAE
Rush Family

Perennial or annual herbs from an erect or horizontal rhizome. Leaves mostly basal, linear, sheathing at base, sheaths usually open, sometimes reduced to sheaths. Flowers bisexual or unisexual (the species dioecious), actinomorphic, small, green, in heads, panicles or corymbs. Perianth of 3 + 3 sepaloid tepals. Androecium of 6 [3] stamens, the pollen in tetrads. Gynoecium of 3 united carpels, unilocular, placentation axile or parietal, style 1, stigmas 3 and brush-like, the ovary superior. Fruit a loculicidal capsule.

9/400; largely of cool temperate and subarctic, damp and wet sites. Two genera are native to the U.S. Of no direct economic importance to man. A few are ornamentals.

Selected Genera:

Juncus (220)- rush

Luzula (70-80)- woodrush; like *Juncus*, but hairy

Recognition Characters:

Grass-like herbs of wet and damp sites with basal, tufted leaves and reduced, inconspicuous 3-merous flowers; fruit a capsule.

$$K^3 \ C^3 \ A^{6[3]} \ \underline{G}^{(3)}$$

CYPERACEAE
Sedge Family

Perennial (rarely annual) herbs of wet and marshy sites. Plants often with a creeping rhizome. Stems usually with solid internodes and often 3-sided. Leaves from basal tufts or cauline and 3-ranked, sheaths usually closed, ligule usually lacking. Flowers bisexual or unisexual (the species usually monoecious), minute, spirally or distichously arranged in tiny spikes. Each flower of the spike is subtended by a small bract, often called a glume. The spike of reduced flowers and subtending glumes or bracts form a unit called the **spikelet.** The spikelets themselves are arranged in various paniculate, umbellate, and spicate inflorescences. Note that in this family and the following one, the usual inflorescence terms do not apply to the disposition of individual flowers, but to the arrangement of spikelets. The perianth, if present at all, is reduced to bristles, hairs, or scales. Androecium of [1] 3 [6] stamens. Gynoecium of 2 or 3 united carpels, unilocular, 1-ovuled, as many style branches as carpels, ovary superior. The pistillate flowers may be surrounded by a sac-like structure, the **perigynium.** Fruit an achene or nutlet, lenticular or 3-sided.

90/4000; widespread, particularly in the cool temperate and subarctic regions. Twenty-four genera are native to the U.S. Of little economic importance; several are cultivated; a few are edible. As treated here, the family includes the Kobresiaceae.

Selected Genera:

A. Flowers unisexual, the gynoecium enclosed by a perigynium

    Subfamily: Caricoideae

        Carex (1100)- sedge

        Kobresia (50

A. Flowers bisexual, if unisexual then the gynoecium not enclosed by a perigynium

    Subfamily: Cyperoideae (Flowers in many-flowered spikelets)

        Cyperus (600+)- sedge, galingale

        Eriophorum (15)- cotton grass

        Scirpus (250)- bulrush, tule

        Eleocharis (150)- spike rush

        Dulichium (1)- *D. arundinaceum* is the 3-way sedge

    Subfamily: Rhynchosporoideae (flowers in few-flowered spikelets)

Rhynchospora (250)- beaked rush

Recognition Characters:

Grass-like herbs with solid, 3-sided stems, 3-ranked leaves, closed sheaths, and reduced wind-pollinated flowers subtended by a single bract. The family is most easily confused with the Gramineae. See treatment of that family for a comparison.

Fig. 97. **Cyperaceae.** a, *Cyperus* spikelet; b, *Carex* showing pistillate flower emerging from the perigynium (indicated by the arrow); c, stem in x-s; d, triangular stem surrounded by closed sheath; e, *Scirpus* flower and bract; f, *Cyperus.* **Juncaceae.** g & h, *Juncus* plant and flower with young capsule. (Drawings b & e after Melchior & Werdermann, 1964).

GRAMINEAE (= POACEAE)
Grass Family

Annual or perennial herbs, rarely tree-like, as in the bamboos. Roots fibrous; rhizomes and stolons frequent. Stems **(culms)** round, often with hollow internodes, swollen nodes. Leaves alternate, simple, 2-ranked, and usually differentiated into a lamina, sheath, and **ligule** (a membrane or series of hairs at the point of junction of the lamina and sheath). Sheath usually open, with the edges meeting or overlapping slightly. Flowers bisexual or unisexual (the species monoecious or dioecious), minute, typically wind-pollinated. Flowers in tiny spikes. Each flower subtended by 2 bracts. This unit (composed of flowers and bracts) is the **spikelet.** It is borne on a pedicel. The central axis of the spikelet is the **rachilla.** Attached to the rachilla in a 2-ranked fashion are bracts, some with flowers in their axils, some without. In the generalized spikelet, the 2 lowermost bracts are sterile; they lack flowers. These bracts, inserted alternately, are the **first** and **second glumes.** Those bracts above the glumes are called **lemmas.** In the axil of most lemmas is a tiny flower, immediately subtended by a second bract called the **palea.** The palea is inserted on the pedicel of an individual flower and not on the rachilla. The palea is partially enclosed by the lemma. These two bracts form a two-bract system enclosing the flower. This subunit of the spikelet is called a **floret.** A spikelet is usually composed of 2 glumes and one or more florets. The glumes, lemmas, and paleas generally have almost microscopic ridges running their length. These strands of vascular tissue are called **nerves.** Their presence and number are of taxonomic importance. The number of nerves on the lemmas and glumes is variable; the palea typically as 2. The glumes and lemmas, but not the palea, may also bear stiff hair-like processes on them called **awns.** The awn may be small and delicate, to several cm long and quite substantial. The awn often arises from the tip of the bract, but it may also come from the back or base.

Spikelets may be round **(terete)** in cross-section or flattened, either **laterally** or **dorsally.** The bracts of a spikelet showing lateral compression are flattened from the sides. Those of a dorsally compressed spikelet are flattened from the backs of the bracts.

Spikelets usually break apart at certain points of weakness. This process is called **disarticulation.** It may be above the glumes and between the florets, so that the glumes remain on the pedicels; or below the glumes, so that the entire spikelet falls from the pedicel.

The flower of the Gramineae lacks a well-developed perianth. It is reduced to 2 or 3 microscopic structures called **lodicules.** These are very sensitive to changes in water pressure and swell, thereby forcing the lemma and palea apart to facilitate wind pollination. Not all grass flowers have lodicules. The androecium is composed of [2] 3 [6] stamens, with comparatively large anthers, a standard feature of anemophilous flowers. The gynoecium is composed of 3 united carpels, one of which is functional. The ovary is unilocular and one-seeded. There are 2 stigmas, elaborately dissected and feather-like. The fruit is typically the caryopsis or grain; rarely an achene or berry.

600/10,000; cosmopolitan. It is the most commonly occurring flowering-plant family, found in practically all habitats and on all continents, including Antarctica. Over 180 genera and almost 1000 species occur in the U.S. The family is of immense economic importance to man. Many are food plants, others are ornamentals, forage crops, weeds, and even building materials in the case of bamboos. Many of these grasses are indicated below.

The Gramineae have been intensively studied in recent years, resulting in some rather extensive changes in concepts of subfamilies and tribes. The newer treatments are complex, based to a large extent upon cytogenetic and anatomical data.

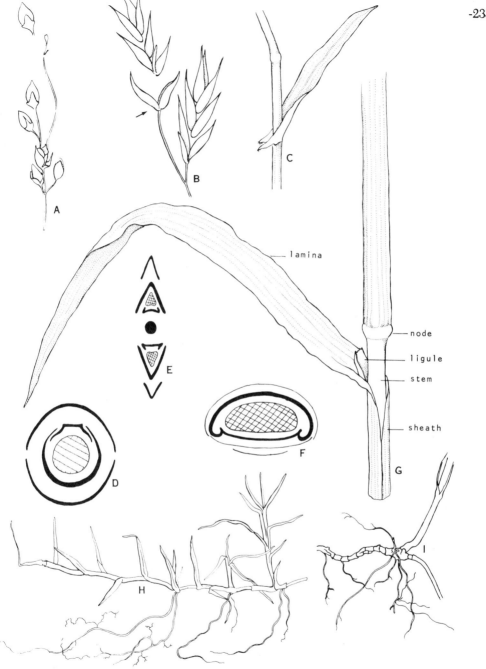

Fig. 98. **Grass morphology.** a, spikelet showing disarticulation below the glumes; b, spikelet showing disarticulation above glumes and between florets (arrow indicates empty glumes); c, leaf base showing auricles; d, terete spikelet; e, spikelet showing lateral compression; f, spikelet showing dorsal compression; g, leaf and stem structure showing lamina, ligule, sheath and swollen node; h, stolon; i, rhizome. (Drawings d-f after Pohl, 1968).

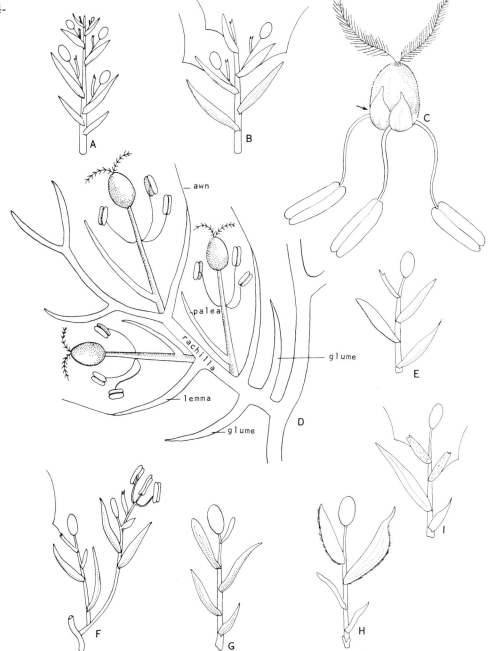

Fig. 99. **Gramineae.** a, *Festuca* spikelet diagram; b, *Avena* spikelet diagram; c, flower (arrow indicates lodicule); d, generalized spikelet; e, *Agrostis* spikelet diagram; f, *Andropogon* spikelet diagram; g, *Panicum* spikelet diagram; h, *Oryza* spikelet diagram; i, *Anthoxanthum* spikelet diagram. (Drawings a & b; e-i after Melchior and Werdermann, 1964).

Selected Genera:

Subfamily: Bambusoideae

Tribe: Arundinarieae

Arundinaria (100)- cane, switchcane, native to the U.S.

Subfamily: Festucoideae (= Pooideae)

Tribe: Festuceae (= Poeae)

Bromus (100)- brome grass

Festuca (100)- fescue grass

Lolium (10)- wild rye

Poa (250)- blue grass

Dactylis (3)- *D. glomerata* is orchard grass

Tribe: Aveneae (including

Agrostideae and Phalarideae)

Avena (10-15)- oats

Holcus (8)- velvet grass

Anthoxanthum (4)- sweet vernal grass

Phalaris (15)- canary grass

Ammophila (4)- beach grass

Agrostis (125)- bent grass

Phleum (10)- *P. pratense* is timothy

Tribe: Triticeae (= Hordeae)

Agropyron (60)- wheat grass

Triticum (14-17)- *T. aestivum* is bread wheat

Secale (5)- *S. cereale* is rye

Elymus (50)- wild rye

Hordeum (25)- *H. vulgare* is barley

Tribe: Meliceae

Melica (60)- onion grass

Glyceria (35)- manna grass

Tribe: Stipeae

Stipa (150)- needle grass

Subfamily: Arundinoideae

Tribe: Arundineae

Arundo (6)- reed grass

Phragmites (2-3)- pampas grass

Cortaderia (15)- pampas grass

Tribe: Aristideae

Aristida (200)- three-awn grass

Subfamily: Oryzoideae

Tribe: Oryzeae (including the Zizanieae)

Oryza (15-20)- *O. sativa* is rice

Leersia (10)- rice cut grass

Zizania (3)- *Z. aquatica* is wild rice

Subfamily: Eragrostoideae

Tribe: Eragrosteae (including the Sporoboleae)

Eragrostis (250)- love grass, stink grass

Eleusine (6)- *E. indica* is goose grass

Sporobolus (100)- dropseed

Muhlenbergia (100)- muhly

Tribe: Chlorideae

Chloris (70)- windmill grass; finger grass

Cynodon (10)- bermuda grass

Bouteloua (50)- side oats grama, grama grass

Buchloë (1)- *B. dactyloides* is buffalo grass

Spartina (16)- cord grass

Subfamily: Panicoideae

Tribe: Paniceae (including the Melinideae)

Panicum (600)- panic grass

Paspalum (400)- dallis grass, bahia

Echinochloa (20)- barnyard grass

Setaria (125)- foxtail

Cenchrus (20)- sandbur

Digitaria (300)- crabgrass

Tribe: Andropogoneae (including the
Maydeae or Tripsaceae)

Saccharum (12)- *S. officinarum* is
sugar cane

Andropogon (113)- blue stem

Schizachyrium (50)- little bluestem

Sorghum (35)- *S. bicolor* is sorghum

Sorghastrum (15)- Indian grass

Tripsacum (7)- gamagrass

Zea (including *Euchlaena*) (3)- *Z. mays* is maize

Table 20. Comparison of the tribes of the Gramineae.

Subfamily-Tribe	Inflorescence	Florets/ Spklt.	Com-pression[1]	Disarticulation	Miscellaneous Comments
Bambusoideae					
Arundinarieae	Panicle	1-x	L	above	Several bamboos cultivated
Festucoideae					
Festuceae	Panicle [raceme]	2-x [1]	L	above	1 or both glumes shorter than lemma
Aveneae	Panicle	1-x	L	above	Glumes as long as lowest lemma
Triticeae	Spike	2-x	L	above	Spikelets usually 1-3 per node
Meliceae	Panicle, raceme	x	L	above or below	Conspicuous lemma nerves
Stipeae	Panicle	1	T	above	Several with long awns
Arundinoideae					
Arundineae	Panicle	2-x	L	above	Pampas grass tribe
Aristideae	Panicle	1	T	above	Awn usually 3-parted
Oryzoideae					
Oryzeae	Panicle	1	L,T	below	Glumes rudimentary or absent
Eragrostoideae					
Eragrosteae	Panicle	1-x	L	above	Lemmas 3 [1]-nerved
Chlorideae	Spikes	1-x	L	above	Often 1 fertile floret
Panicoideae					
Paniceae	Panicle	2	D	below	Fertile floret hard or leathery
Andropogoneae	Panicle	2	D	below	Fertile floret soft

1) L= lateral, T= terete, D= dorsal

Recognition Characters:

Herbs with linear 2-ranked leaves and round hollow culms, ligules, open sheath; flowers greatly reduced, enclosed in two bracts, arranged in spikelets. The family is most easily confused with the Cyperaceae. The two families may be separated as follows:

Table 21. Comparison of the Cyperaceae and Gramineae.

character	Cyperaceae	Gramineae
stems	internodes solid stems 3-sided	internodes hollow stems round
leaves	3-ranked sheaths closed eligulate	2-ranked sheaths open ligulate
flowers	subtended by 1 bract lodicules absent	subtended by 2 bracts lodicules present
fruit	achene	caryopsis

SPARGANIACEAE
Bur-reed Family

Perennial, rhizomatous aquatic herbs. Leaves alternate, simple, ± linear, erect or floating. Flowers unisexual (the species monoecious); hypogynous; in separate globose heads, the staminate heads above the pistillate ones. Staminate flowers with 3-6 membranous sepaloid tepals and 3 or more stamens. Pistillate flowers with 3-6 sepaloid tepals and a unicarpellate, unilocular gynoecium. Fruit variously interpreted as nutlike or drupaceous.

1/20; primarily in the cooler regions of the north temperate zone, Australia, and New Zealand; absent from South Africa and South America. Several species are native to the U.S. Of no direct economic importance to man.

Genus:

Sparganium (20)- bur-reed

Recognition Characters:

Perennial aquatic herbs with ± linear leaves. Flowers unisexual, 3-merous, in separate globose heads, the staminate above the pistillate.

$$K^{3\text{-}6}C^0A^{3\text{-}6}G^0 \quad \text{(staminate flower)}$$

$$K^{3\text{-}6}C^0A^0G^1 \quad \text{(pistillate flower)}$$

Fig. 100. **Sparganiaceae.** *Sparganium.*

TYPHACEAE
Cattail Family

Perennial herbs of marshes and open water. Rhizome creeping. Leaves mostly basal, linear, erect, 2-ranked. Flowers unisexual, the species monoecious, borne in a tightly compacted terminal spadix, the staminate flowers above, the pistillate below. Perianth reduced to a series of slender jointed threads, bristles, or scales. Staminate flowers with 2-5 stamens, free or monadelphous, pollen in tetrads. Pistillate flowers with a stipitate, unicarpellate, unilocular ovary, the stipe bearing numerous silky hairs. Fruit an achene or nutlet.

1/15; temperate and tropical. *Typha* is widespread in the U.S. Of little economic importance. A few species are ornamentals and several have been used in matting.

Genus:

Typha (15)- cattail

Recognition Characters:

Large marsh herbs with 2-ranked linear leaves and a brownish compact spike of unisexual flowers.

K ^{x} C ^{0} A $^{2\text{-}5}$ G ^{0} (staminate flower)

K ^{x} C ^{0} A ^{0} $\underline{G}$ ^{1} (pistillate flower)

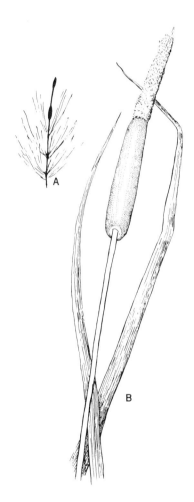

Fig. 101. **Typhaceae.** *Typha* pistillate flower and inflorescences.

BROMELIACEAE
Pineapple Family

Epiphytic or terrestrial herbs. Leaves alternate, usually basal and rosette forming, often rigid and spine-toothed, and basally colored. Flowers usually bisexual, actinomorphic, 3-merous, hypogynous or epigynous, often subtended by brightly colored spathaceous bracts. Calyx of 3 free sepals. Corolla of 3 free or connate petals, usually brightly colored. Androecium of 6 stamens, often epipetalous. Gynoecium of 3 united carpels, 3-locular, placentation axile, ovary superior or inferior. Fruit a berry, capsule, or multiple *(Ananas)*.

50/2000; primarily of the New World tropics. A single species, *Pitcairnia feliciana*, may be native to Africa. Four genera are native to the U.S., with the family most common in the Southeast and Gulf Coast region. Of moderate economic importance as a source of food, ornamentals, and fibers.

Selected Genera:

Subfamily: Pitcairnioideae (mostly terrestrial; ovary superior; fruit a capsule)

Puya (140)- South American; some plants quite tall

Hechtia (41)- native to Texas

Pitcairnia (260)- popular ornamentals

Subfamily: Tillandsioideae (mostly epiphytic; ovary superior; fruit a capsule; seeds appendaged)

Tillandsia (350)- *T. usneoides* is Spanish moss

Guzmania (120)- native to Florida

Catopsis (25)- air plant; native to Florida

Subfamily: Bromelioideae (epiphytic; ovary inferior; fruit a berry or multiple; no appendages on seeds)

Aechmea (150)- air plant; commonly cultivated

Billbergia (50)- vase plant; commonly cultivated

Ananas (5-6)- *A. comosus* is the pineapple

Recognition Characters:

Epiphytic or terrestrial herbs with rosettes of often spiny leaves and brightly colored 3-merous flowers in conspicuous inflorescences subtended by brightly colored bracts.

$$K \; 3 \; C \; 3 \; A \; 6 \; \overline{\underline{G}} \; (3)$$

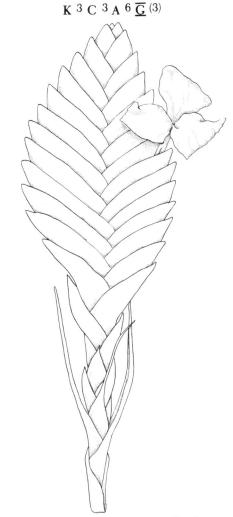

Fig. 102. **Bromeliaceae.** Inflorescence of *Tillandsia*.

MUSACEAE
Banana Family

Tree-like perennial herbs from branched rhizomes. Leaves large, simple, (lamina is soon frayed by the wind), spirally arranged, pinnately veined, the bases overlapping to form what appears to be an aerial stem. Flowers usually unisexual (the species monoecious), zygomorphic, epigynous, in pendant racemes subtended by brightly colored spathaceous bracts. Perianth biseriate, 3 + 3 or 5 united and 1 free. Staminate flowers with 5 fertile stamens and 1 staminodium. Pistillate flowers with a gynoecium of 3 united carpels, 3-locular, and epigynous. Fruit a leathery berry. The commercial banana is usually seedless.

2/70; Africa, Asia, and Australia. None is native to the U.S. Of moderate economic importance because of the banana, platano, and Manila hemp. As treated here, the family does not include several genera now assigned to the Strelitziaceae and Heliconiaceae.

Genera:

Musa (60)- *M. paradisiaca* is the banana and platano

Ensete (10)- Abyssinian banana; widely cultivated

Recognition Characters:

Large tree-like herbs with pinnately veined leaves. Flowers unisexual, zygomorphic, 3-merous, and epigynous. Fruit a leathery berry, ± resembling the familiar banana.

$$K^3 C^{2+1} A^{5+1} G^0 \quad \text{(staminate flower)}$$

$$K^3 C^{2+1} A^0 \overline{G}^{(3)} \quad \text{(pistillate flower)}$$

Fig. 103. **Musaceae.** a, *Musa* (arrows indicate staminate and pistillate portions of the inflorescence). **Zingiberaceae.** b, *Hedychium.* **Heliconiaceae.** c & d, *Heliconia* inflorescence and flower detail. **Strelitziaceae.** e, *Strelitzia* inflorescence showing two flowers having emerged from a large subtending bract (indicated by arrow).

Liliopside

Order: Family	Common Name	Genera/Species	Distribution	Form	K
Arecales					
Palmae	Palm	236/3400	Trop. & subtrop.	T	3
Arales					
Araceae	Philodendron	115/2000	Chiefly trop.	HS	0,4-6
Lemnaceae	Duckweed	4/25	Temp. & trop.	H	0

Arecidae

C	A	G	Fruit Type(s)	Miscellaneous Comments
3	6	(3)	ber, drp	Large leaves: conspicuous inflorescences
0	6	(2-3)	ber	Leaves often leathery and pinnately veined
0	1	1	utr	Free-floating aquatics; smallest flw. plants

The subclass consists of four orders, five families, and 6400 species. More than one - half are palms. The flowers are typically small, numerous, and subtended by a spathe. Most have broad petiolate leaves with pinnate venation.

The Arecales are typically trees with a simple trunk and a terminal cluster of leaves. The Arales consist of the Lemnaceae, a family of minute free - floating aquatics; and the Araceae, a family of terrestrial and epiphytic herbs with leathery pinnately-veined leaves.

PALMAE (= ARECACEAE)
Palm Family

Trees, shrubs, and vines. Stems variable, ranging from large unbranched trunks through slender and flexuous forms, to very short and seemingly absent. Leaves (fronds) often large (to 15 m in length), alternate, petiolate, the bases usually sheathing. The leaves may be clustered into a crown at the apex of the trunk (the arborescent species) or scattered along the trunk. The leaves simple and pinnately veined (the condition in a few primitive genera); simple (or compound depending upon interpretation) and flabellate (the "fan palms"); or compound and pinnate (the "feather palms"). The compound condition arises by a splitting of a continuous lamina. Flowers bisexual or unisexual (the species monoecious or dioecious), small, 3-merous, hypogynous, in conspicuous paniculate inflorescences, the inflorescence subtended by a spathe. Perianth biseriate, 3 + 3, sepals and petals free or connate. Androecium of 6 stamens in two series, rarely numerous. Gynoecium of 3 distinct or basally united carpels, 1-3 locular, 1 ovule/locule, ovary superior. Fruit a berry or drupe; exocarp often fibrous.

236/3400; primarily tropical and subtropical. Eight or nine genera are native to the U.S., and almost a hundred are in cultivation in the warmer parts of the country. The family is of immense economic importance, probably second only to the Gramineae. Products include edible fruits, oils, waxes, building materials, and numerous ornamentals. The family name Palmaceae, which appears in several references, is incorrect.

Selected Genera:

Subfamily: Coryphoideae (G^3; separating after fertilization; berry)

Phoenix (13)- *Ph. dactylifera* is the date palm

Sabal (22)- palmetto, cabbage palm

Washingtonia (2)- California fan palm

Copernicia (29)- *C. prunifera* is the source of carnauba wax

Subfamily: Borassoideae (G$^{(3)}$; fully united; drupe)

Lodoicea (1)- *L. maldivica* is the coco de mer or double coconut, reputed to be the largest "seed"

Subfamily: Lepidocaryoideae (G$^{(3)}$; fully united; fruit scaly)

Calamus (340)- *C. rotang* is the source of rattan cane

Metroxylon (28)- sago palm

Subfamily: Arecoideae (G$^{(3)}$; fruit smooth)

Cocos (1)- *C. nucifera* is the coconut palm

Roystonea (16)- royal palm; widely cultivated

Areca (88)- *A. catechu* is the betel nut palm

Bactris (210)- spiny club palm, pejibaye

Elaeis (8)- *E. guineensis* is the African oil palm.

Recognition Characters:

Often medium to tall trees with well-developed unbranched trunks, large fan-shaped or pinnately compound leaves with sheathing bases and a conspicuous spathaceous paniculate inflorescence of small flowers.

K 3 C 3 A 6 $\underline{G}$ 3 or $\underline{G}$ (3)

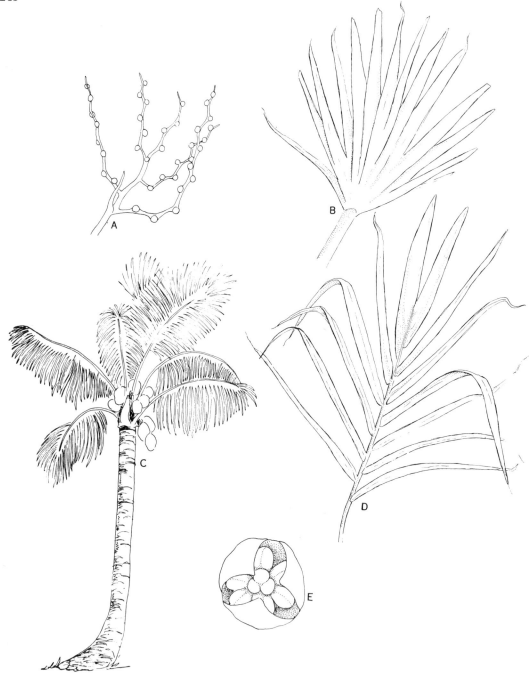

Fig. 104. **Palmae.** a, young inflorescence branches; b, fan palm leaf; c, *Cocos;* d, feather palm leaf; e, flower.

ARACEAE
Philodendron Family

Mostly terrestrial herbs, rarely woody and climbing. Rhizomatous or tuberous. Sap watery or milky. Leaves alternate, simple or compound, petiolate, base sheathing; pinnately, palmately, or parallel veined. Flowers bisexual or unisexual (the species monoecious for the most part), small, actinomorphic, hypogynous, in a spadix often subtended by a large, showy spathe. Perianth of 4-6 free or connate segments; usually present in bisexual flowers, typically absent in unisexual ones. Androecium typically of 6 stamens, opening by pores or slits; staminodia sometimes present. Gynoecium of 2 or 3 united carpels [1-9], quite variable, 1- many locules, ovary usually superior (sometimes embedded in the spadix axis), placentation variable. Fruit a berry.

115/2000; primarily tropical, but many species also occur in temperate regions. Eight genera are native to the U.S. The family is of some economic importance on a world wide basis as a source of starchy foods; many are cultivated as ornamentals. The plant tissues often contain needle-like crystals of calcium oxalate which can cause mechanical injury to delicate linings of the mouth and throat.

Selected Genera:

Subfamily: Pothoideae
  Anthurium (500)- anthurium, tailflower
  Acorus (2)- *A. calamus* is the sweetflag

Subfamily: Monsteroideae
  Monstera (30)- window leaf, ceriman
  Spathiphyllum (36)- spathe flower

Subfamily: Calloideae
  Calla (1)- *C. palustris* is the water arum
  Lysichiton (1)- *L. americanum* is the western skunk cabbage
  Symplocarpus (1)- *S. foetidus* is the eastern skunk cabbage

Subfamily: Lasiodeae
  Amorphophallus (80)- devil's tongue, leopard palm; *A. titanum* called the world's largest "flower"

Subfamily: Philodendroideae
  Philodendron (250)- philodendron
  Zantedeschia (8)- calla, calla lily
  Aglaonema (45)- Chinese evergreen
  Dieffenbachia (30)- dumb cane; popular house plant

Subfamily: Colocasioideae
  Colocasia (6)- *C. esculenta* is taro or dasheen
  Alocasia (60)- elephant's ear
  Xanthosoma (40)- yautia, its tubers edible

Subfamily: Aroideae
  Arum (12)- cuckoo-pint
  Arisaema (130)- Indian turnip, jack-in-the-pulpit

Subfamily: Pistioideae
  Pistia (1)- *P. stratiotes* is water lettuce

Recognition Characters:

(Ours) herbs, typically of wet sites, with leathery, often pinnately veined leaves, and a spadix and spathe.

LEMNACEAE
Duckweed Family

Small to minute free-floating and submerged perennial herbs of freshwater habitats. Roots simple and thread-like or completely lacking. Plant body leafless, reduced to an oval or globose thallus or frond. Flowers unisexual (the species monoecious) and greatly reduced. Perianth absent, a membranous spathe initially enclosing the flower. Staminate flowers of 1 or rarely 2 stamens. Pistillate flowers of a single unicarpellate, unilocular gynoecium. Fruit a utricle.

4/25; temperate and tropical. All four genera occur in the U.S. Other than their use as aquarium ornamentals, the family is of no direct economic importance to man.

Genera:

Lemna (7-10)- duckweed

Spirodela (3)- greater duckweed

Wolffiella (2)- mud-midget, bogmat

Wolffia (12-14)- water meal; *W. arrhiza* is the smallest flowering plant

Recognition Characters:

Small to minute (larger plants only a few mm in diameter) free-floating aquatics of freshwater sites. Flowering seemingly rare; asexual reproduction common.

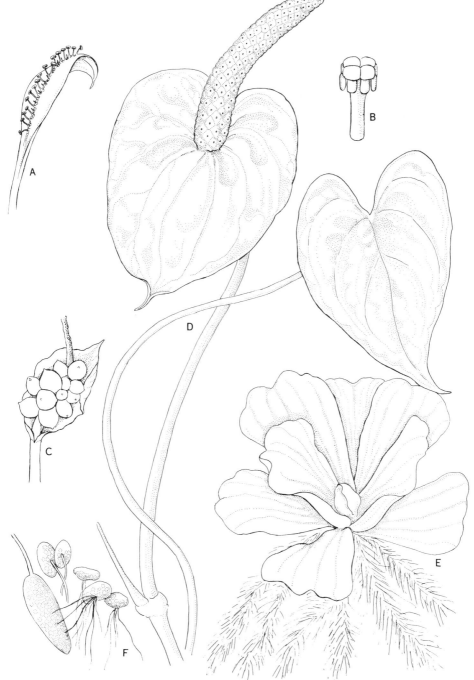

Fig. 105. **Araceae.** a, *Spathicarpa;* b, staminate flower; c, *Aglaonema* fruits; d, *Anthurium* spadix and spathe; e, *Pistia.* **Lemnaceae.** f, *Lemna* and *Spirodela.*

Liliopsida

Order: Family	Common Name	Genera/Species	Distribution	Form	K
Liliales					
Pontederiaceae	Pickerel weed	7/30	Trop. & subtrop.	H	3
Liliaceae	Lily	40/6510	Cosmopolitan	HS	3
Iridaceae	Iris	70/1500	Temp. & trop.	H	3
Haemodoraceae	Bloodwort	14/75	Austr., S. Am., S. Afr	H	3
Dioscoreaceae	Yam	11/650	Trop. & warm temp.	VS	3
Burmanniales					
Burmanniaceae	Burmannia	17/125	Trop. & subtrop.	H	3
Orchidaceae	Orchid	6-700/20,000	Cosmo.	H	3

.iliidae

C	A	G	Fruit Type(s)	Miscellaneous Comments
3	6	(3)	cap, nut	Emergent aquatics
3	6[3]	(3)	cap, ber	Here treated in broad sense
3	3	(3)	cap	Leaves typically equitant
3	6	(3)	cap	*Lachnanthes* along Atlantic coastal plain
3	6	(3)	cap, ber	Only *Dioscorea* is native to U.S.
3	3	(3)	cap	Three genera in Eastern U.S.
2+1z	1-2	(3)	cap	Lip often quite elaborate

The Liliidae consist of two orders, fifteen families, and almost 28,000 species. It is the largest subclass of the Liliopsida. Two - thirds of them are orchids. Plants of this group typically are herbs whose flowers have a petaloid perianth and a syncarpous gynoecium.

In the Liliales the embryo is well - developed and endosperm is present. I have departed from Cronquist in his treatment of the order by merging the Agavaceae and Smilacaceae with the Liliaceae. In the Orchidales the embryo is relatively undifferentiated and endosperm essentially absent.

PONTEDERIACEAE
Pickerel-Weed Family

Perennial aquatic herbs of freshwater sites, floating or attached. Rootstock thick. Leaves opposite or whorled, bases sheathing. Flowers bisexual, usually actinomorphic, 3-merous, in racemes or panicles, subtended by a spathaceous sheath or petiole. Perianth biseriate, 3 + 3, petaloid, free or connate. Andorecium of 6 [3 or 1] epipetalous stamens. Gynoecium of 3 united carpels, 3-locular, ovary superior. Fruit a capsule or nut.

7/30; primarily tropical and subtropical. Three genera are indigenous to the U.S. Some are ornamentals and *Eichhornia* has become a major weed of waterways.

Selected Genera:

Pontederia (3-4)- pickerel weed

Heteranthera (10)- mud plantain

Eichhornia (6)- *E. crassipes* is the water hyacinth

Recognition Characters:

Freshwater aquatics, either rooted or free - floating. Flowers 3-merous, the perianth petaloid.

$$K^3 C^3 A^6 \underline{G}^{(3)}$$

Fig. 106. **Pontederiaceae.** *Eichhornia*.

LILIACEAE
Lily Family

Mostly perennial herbs, sometimes quite robust, from bulbs, rhizomes, corms, sometimes tuberous. Leaves alternate, simple, linear (broader in the Smilacoideae and Asparagoideae); few and basal in the Amaryllidoideae; fleshy and fibrous in the Agavoideae. Flowers bisexual, less frequently unisexual (the species usually dioecious); actinomorphic, rarely zygomorphic; typically showy, borne in various racemose inflorescences; 3-merous, the perianth 3 + 3, petaloid, free or united into a tube. Corona present or absent. Androecium of 6 [3] stamens. Gynoecium of 3 united carpels, 3-locular, placentation axile, the ovary superior, half-inferior, or inferior. Fruit a loculicidal or septicidal capsule or berry.

40/6510; cosmopolitan. About 75 genera are native to the U.S. The family is the source of many handsome ornamentals, several important fibers, fermented and distilled beverages, and steroidal compounds. Most recent authors have accepted the view that the Liliaceae are closely related to the Amaryllidaceae and Agavaceae. The delimitation of these families, along with numerous transitional ones, seems entirely arbitrary to me. The Liliaceae and Amaryllidaceae have been separated by the nature of the inflorescence or by ovary position. I concur with Cronquist that neither of these features is useful. The Agavaceae were once thought to be more objectively defined by their haploid chromosome complement of 5 large chromosomes and 25 small ones. Otherwise the Agavaceae combine floral characters of the Liliaceae and Amaryllidaceae. More recent studies have suggested that the peculiar 5 + 25 chromosome pattern is not unique to the family, nor is it found within all species of the Agavaceae. Therefore, separation of the family from basic liliaceous stock is difficult. As treated here, the family includes the Abamaceae, Agavaceae, Agapanthaceae, Alliaceae, Aloaceae, Alstroemeriaceae, Amaryllidaceae, Anthericaceae, Aphyllanthaceae, Asparagaceae, Asphodelaceae, Aspidistraceae, Asteliaceae, Balbocodiaceae, Bractillaceae, Brunsvigiaceae, Calochortaceae, Cepaceae, Colchicaceae, Compsomaceae, Convallariaceae, Cymbanthaceae, Cyrtanthaceae, Dasypogonaceae, Dianellaceae, Dracaenaceae, Eriospermaceae, Eucomidaceae, Fritillariaceae, Funkiaceae, Galanthaceae, Gethyllidaceae, Gilliesiaceae, Haemanthaceae, Helionadaceae, Hemeriocallidaceae, Herreriaceae, Hyacinthaceae, Ixiolirionaceae, Johnsoniaceae, Lachenaliaceae, Lapageriaceae, Leucojaceae, Lophiolaceae, Luzuriaceae, Melanthiaceae, Miyoshiaceae, Narcissiaceae, Nartheciaceae, Nolinaceae, Ophiopogonaceae, Oporanthaceae, Ornithogalaceae, Pancratiaceae, Paridaceae, Peliosanthaceae, Petermanniaceae, Petrosaviaceae, Philesiaceae, Phormiaceae, Platymetraceae, Polygonataceae, Protoliriaceae, Ruscaceae, Sansevieriaceae, Sarmentaceae, Scillaceae, Smilacaceae, Sturmariaceae, Themidaceae, Trilliaceae, Tulbaghiaceae, Tulipaceae, Uvulariaceae, Veratraceae, Yuccaceae, and Zephyranthaceae.

Selected Genera:

Group A: Genera associated with the traditional Liliaceae

Subfamily: Melanthioideae (= Melanthiaceae)

Zigadenus (= *Zygadenus*) (9)- death camas; toxic

Veratrum (48)- false hellebore, corn lily; toxic

Subfamily: Asphodeloideae

Chlorophytum (100)- St. Bernard's lily, spider plant

Hosta (40)- plantain lily

Hemerocallis (16)- day lily

Aloë (250)- aloe, very popular ornamentals

Subfamily: Wurmbaeoideae

Colchicum (65)- autumn crocus, the source of colchicine

Gloriosa (6)- G. superba is the glory lily

Subfamily: Lilioideae

Calochortus (40)- mariposa lily

Erythronium (20)- fawn lily, dogtooth violet

Tulipa (60)- tulip

Fritillaria (100)- fritillary

Lilium (75)- lily

Subfamily: Alstroemerioideae (= Alstroemeriaceae)

Alstroemeria (60)- lily of the Incas

Subfamily: Scilloideae (= Scillaceae)

Scilla (100)- squill

Camassia (4)- camas, quamash

Ornithogalum (100)- star-of-Bethlehem

Hyacinthus (1)- H. orientalis is the hyacinth

Subfamily: Smilacoideae (= Smilacaceae)

Smilax (300)- green briar, sarsaparilla

Subfamily: Asparagoideae

Convallaria (1)- C. majalis is the lily-of-the-valley

Smilacina (25)- false solomon's seal

Maianthemum (3)- false lily-of-the-valley

Polygonatum (30)- Solomon's seal

Trillium (30)- trillium, wake robin

Asparagus (300)- asparagus

Ruscus (2-3)- butcher's broom, an ornamental with cladophylls

Group B: Genera variously placed in the Liliaceae or Amaryllidaceae

Subfamily: Allioideae (= Alliaceae)

Agapanthus (9)- lily-of-the-Nile

Allium (300)- onion, garlic, chives, leek

Brodiaea (10)- brodiaea

Dichelostemma (6)- blue dicks, wild hyacinth

Triteleia (14)- triteleia, wild hyacinth

Group C: Genera associated with the traditional Amaryllidaceae

Subfamily: Amarylloidideae (= Amaryllidaceae)

Crinum (150)- crinum lily, spider lily

Amaryllis (1)- A. belladonna is the naked lady

Hippeastrum (75)- amaryllis of cultivation

Haemanthus (80)- blood lily

Galanthus (12)- snow drops

Hymenocallis (36)- spider lily

Narcissus (22)- narcissus, daffodil

Group D: Genera associated with the Agavaceae

Subfamily: Agavoideae (= Agavaceae)

Agave (300)- century plant, maguey, sisal hemp; the source of tequila and steroidal hormones

Yucca (35)- yucca, Spanish bayonet, Our Lord's candle

Cordyline (20)- dracaena

Nolina (25)- bear grass, sacahuista

Dasylirion (25)- sotol

Sansevieria (50)- bowstring hemp

Recognition Characters:

Perennial herbs from bulbs, rhizomes or corms. Leaves linear, sometimes sword-shaped and fibrous. Flowers with a conspicuous 6-parted petaloid perianth, 6 [3] stamens, and a superior or inferior tricarpellate ovary. Fruit a capsule or berry.

$$K \ 3 \ C \ 3 \ A \ 6 \ \overline{\underline{G}} \ (3)$$

Fig. 107. **Liliaceae.** a & b, *Ornithogalum* raceme and plant; c, *Smilax*; d, *Agave*; e & f, *Hippeastrum* fruit with withered perianth and longitudinal section of flower; g, *Haemanthus*; h, *Ornithogalum* flower.

IRIDACEAE
Iris Family

Perennial herbs from rhizomes, bulbs, or corms. Leaves alternate, simple, basal, sheathing, equitant. Flowers bisexual, actinomorphic or zygomorphic, 3-merous, epigynous, in terminal cymose inflorescences. Perianth biseriate, 3+3, petaloid, united below into a straight or curved tube. Androecium of 3 stamens. Gynoecium of 3 united carpels, 3-locular, placentation axile, ovary inferior, style tifid; petaloid in *Iris*. Fruit a loculicidal capsule.

70/1500; temperate and tropical. Chief centers of distribution in South Africa and tropical America. Five genera are native to the U.S. Of importance as a source of ornamentals and saffron dye.

Selected Genera:

Iris (200)- iris, flag, fleur-de-lis

Sisyrinchium (80)- blue-eyed grass

Tigridia (15)- tiger flower

Freesia (3)- freesia, a common ornamental

Crocus (80)- *C. sativus* is the source of saffron

Gladiolus (250)- glads, sword lily

Watsonia (60)- bugle lily

Crocosmia (5)- montbretia

Recognition Characters:

Herbs with equitant leaves, petaloid perianth, 3 stamens, (petaloid styles in *Iris!*), and a 3-carpellate inferior ovary.

$$K \underline{3} C \underline{3} A 3 \overline{G} (3)$$

Fig. 108. **Iridaceae.** a, *Iris* flower with style pulled back to show stamen; b, leaves and flower of *Iris* (arrow indicates petaloid style) c, *Iris* capsules.

ORCHIDACEAE
Orchid Family

Perennial herbs; terrestrial, epiphytic, sometimes saprophytic. Most, if not all, species have a mycorrhizal relationship. Leaves alternate and often distichous (rarely opposite or whorled), simple, sometimes reduced to scales. Flowers bisexual, rarely unisexual (the species monoecious or dioecious), zygomorphic, (rarely actinomorphic), bracteate, 3-merous, epigynous, in racemes, spikes, or panicles. Calyx of 3 sepals, either green and sepaloid; or brightly-colored and petaloid. Corolla of 3 petals, the upper one appearing as the lower because of a 180° twist in the flower (resupination). The lower petal is often elaborately modified into a lip or labellum. It may be a rather simple structure or heavily modified into a sac or spur. The androecium is composed of 1 or 2 stamens, the anther typically appearing as a cap - like structure. The pollen is usually agglutinated into waxy or mealy masses (pollinia), less frequently in tetrads or separate grains. The tip of a pollinium may have a sterile portion, the caudicle. The gynoecium is composed of 3 united carpels, its placentation parietal, the ovary inferior. The stigmas, styles, and the androecium are combined to form a complex structure called a column or gynandrium. The stigma often appears as a shallow depression on the inner side of the column. All 3 stigmas may be fertile, but more commonly the 2 lateral ones are fertile and the third modified into a small sterile outgrowth, the rostellum. A portion of the rostellum is sometimes further modified into a sticky disc (viscidium) to which the pollinia are attached. Embryo minute, undifferentiated; endosperm lacking.

The Orchidaceae are the largest family of flowering plants, being composed of 600-700 genera and 20,000 species. The family is widespread, abundant in the tropics, where they are usually epiphytic, and becoming increasingly less common in cool temperate and subarctic regions. Twenty-one genera occur in the northeastern U.S.; 13 in the Pacific Northwest; 40 are native or naturalized in Florida; none is native to Alaska. The family is the source of numerous outstanding ornamentals and vanilla flavoring. Hybrids are common in cultivation. As treated here, the family includes the Cypripediaceae, Apostasiaceae, and the Vanillaceae.

Selected Genera:*

A. Fertile anther 2; pollen in tetrads
   Subfamily: Cypripedioideae
      Cypripedium (50)- lady slipper orchid
      Paphiopedilum (50)- lady slipper orchid

A. Fertile anther 1; pollen in pollinial masses
   Subfamily: Neottioideae (pollinia 2-4, soft and mealy)
      Epipactis (20)- stream orchis
      Listera (30)- twayblade
      Goodyera (100)- rattlesnake-plantain
      Spiranthes (35)- ladies-tresses
   Subfamily: Orchidoideae (pollinia in soft masses; caudicles arising from the base of the pollinia)
      Habenaria (450)- orchis
      Ophrys (21)- bee orchid, fly orchid
      Orchis (50)- *O. militaris* is the military orchid
   Subfamily: Epidendroideae (pollinia 2-12, typically hard and waxy; without caudicles, or these terminal)
      Vanilla (100)- *V. planifolia* is the source of vanilla
      Calopogon (4)- grass-pink orchid
      Cattleya (65)- cattleya, popular with florists
      Epidendrum (800)- buttonhole orchid

Corallorhiza (15)- coral root

Vanda (60)- vanda

Oncidium (400)- dancing-lady orchid

*Subfamilies after Van der Pijl, L. and C.H. Dodson. 1966. Orchid flowers, their pollination and evolution. Fairchild Tropical Garden and the University of Miami Press. Coral Gables. pp. 175-178.

Recognition Characters:

Perennial herbs, often epiphytic, with 2-ranked leaves, 3-merous zygomorphic flowers with a labellum, pollinia, gynandrium, and an inferior ovary.

$$K3Cz2+1\underline{A}1\text{-}2\overline{G}(3)$$

Fig. 109. **Orchidaceae.** a, generalized flower (arrow indicates anther cap and gynandrium); b, insect engaging gynandrium; c, anther cap partially dislodged (arrow indicates pollinium); d, generalized plant; e, gynandrium as viewed from below; f, *Vanda;* g, *Cypripedium* showing pouch-like labellum; h, *Paphiopedilum* showing bucket-like labellum. (Drawing a after Bailey, 1949; e, after Wood, 1974).

7- Literature for the Identification of Vascular Plants

I have attempted in the references below to survey the current regional, state, and county floras of the continental United States. In addition, certain specialized monographs and treatments of cultivated plants are included. I have not cited older floras, now only of historic interest, in those states where recent floras have been prepared. Unfortunately, a few states still lack a modern flora. The map showing ranges of the regional floras should assist in these cases.

The references in this chapter are, for the most part, technical treatments. More popular guides designed primarily for the laymen have been omitted. I have included a few "picture books" because of the quality of the photographs and the rigor of the text.

REGIONAL FLORAS AND SPECIALIZED TREATMENTS

Albee, L. R. (Compiler). 1971. National list of scientific plant names. U. S. Department of Agriculture, Soil Conservation Service. Lincoln, Neb. 281 pp.

Abrams, L. 1923-1960. An illustrated flora of the Pacific states. 4 vols. Vol. 4 by Roxana Ferris. Stanford University Press.

Batson, W. T. 1972. A guide to the genera of native and commonly introduced ferns and seed plants of the southeastern United States excluding Peninsular Florida. Publ. by the author. Univ. of S. Carolina. Columbia.

Benson, L. & R. A. Darrow. 1954. The trees and shrubs of the southwestern deserts. Univ. of Arizona and Univ. of New Mexico Presses. Tucson and Albuquerque. 437 pp.

Blackburn, B. 1952. Trees and shrubs in eastern North America. Oxford Univ. Press. New York. 358 pp.

Brockman, C. F. 1968. Trees of North America. Golden Press. New York. 280 pp.

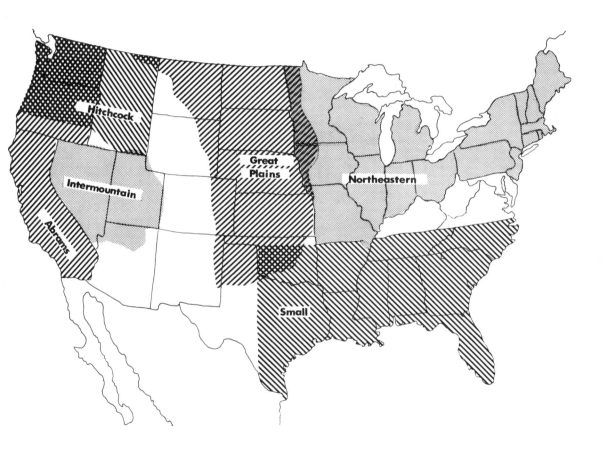

Fig. 110. Map showing the coverage of selected regional U.S. floras.

Case, F. W. 1964. Orchids of the western Great Lakes region. Cranbrook Inst. of Science. Bull. No. 48. Bloomfield Hills, Mich.

Chapman, A. W. 1897. Flora of the southern United States. Amer. Book Co. New York. 655 pp.

Clovis, J. F. et al. 1972. Common vascular plants of the mid-Appalachian region. West Virginia Book Exchange. Morgantown. 306 pp.

Cobb, B. 1956. A field guide to the ferns. Houghton-Mifflin. Boston. 281 pp.

Coker, W. C. & H. R. Totten. 1945. Trees of the southeastern states … 3rd ed. Univ. of North Carolina Press. Chapel Hill. 419 pp.

Correll, D. S. 1950. Native orchids of North America (north of Mexico). Chronica Botanica Co. Waltham, Mass. 399 pp

Correll, D. S. & H. B. Correll. 1972. Aquatic and wetland plants of the Southwestern United States. U. S. Environmental Protection Agency. 1777 pp. (Reprinted by Stanford Univ. Press)

Craighead, J.J., F.C. Craighead, & R.J. Davis. 1963. A field guide to Rocky Mountain wildflowers. Houghton-Mifflin Co. Boston. 277 pp.

Cronquist, A. et al. 1972-. Intermountain flora. Vascular plants of the Intermountain West, U.S.A. To date the first volume of a projected six volumes has been published. Hafner Publ. Co. for the New York Bot. Garden.

Duncan, W. H. 1975. Woody vines of the Southeastern United States. Univ. of Georgia. Athens. 76 pp.

Duncan, W. H. & L. E. Foote. 1975. Wildflowers of the Southeastern United States. Univ. of Georgia Press. Athens. 296 pp.

Ellis, W. H., E. Wofford, & E. W. Chester. 1971. A preliminary checklist of the flowering plants of the Land Between the Lakes. Castanea 36(4):229-246.

Enari, L. 1956. Plants of the Pacific northwest. Binfords & Mort. Portland. 315 pp.

Eyles, D. E. & J. L. Robertson, Jr. 1963. A guide and key to the Aquatic Plants of the southeastern United States. U. .S. Dept. of the Interior. Fish and Wildlife Service. Bureau of Sport Fisheries and Wildlife Circular 158. Washington, D. C.

Fassett, N. C. 1957. A manual of aquatic plants. Revised by E. C. Ogden. Univ. of Wisconsin Press. Madison. 405 pp.

Fernald, M. L. 1950. Gray's manual of botany. 8th ed. American Book Co. New York. 1632 pp.

Gilkey, H. M. 1957. Weeds of the Pacific northwest. Oregon State College. Corvallis. 441 pp.

Gilkey, H. M. & L. R. Dennis. 1967. Handbook of northwestern plants. Oregon State Univ. Bookstores, Inc. Corvallis. 505 pp.

Gleason, H. A. 1963. New Britton and Brown illustrated flora of the northeastern United States and adjacent Canada. 3 vols. Hafner Publ. Co. New York.

Gleason, H. A. & A. C. Cronquist. 1963. Manual of the vascular plants of the northeastern United States and adjacent Canada. Van Nostrand Co. Princeton, N. J. 810 pp.

Gould, F. W. 1951. Grasses of the southwestern United States. Univ. of Arizona Biol. Sci. Bull. No. 7. Tucson. 352 pp.

Great Plains Flora Association. 1976. Atlas of the flora of the Great Plains. Iowa State University Press. Ames. In press.

Harrar, E. S. & J. G. Harrar. 1962. Guide to southern trees. 2nd ed. Dover. New York. 709 pp.

Haskins, L. L. 1967. Wild flowers of the Pacific coast. Binfords & Mort, Publ. Portland. 406 pp.

Hermann, F. J. 1946. A checklist of plants in the Washington-Baltimore area. 2nd ed. Issued by Conf. on District Flora. 130 pp.

Hitchcock, A. S. 1951. Manual of the grasses of the United States. 2nd ed. revised by Agnes Chase. Misc. Publ. No. 200 of the U. S. D. A. 1051 pp.

Hitchcock, C. L. & A. Cronquist. 1973. Flora of the Pacific Northwest: An illustrated manual. University of Washington Press. Seattle. 730 pp.

Hitchcock, C. L. et al. 1955-1969. Vascular plants of the Pacific northwest. 5 vols. University of Wash. Press. Seattle.

Hoffmann, H. L. 1964. Checklist of vascular plants of the Great Smoky Mountains. Castanea 29:1-45.

Holmgren, A. H. & J. L. Reveal. 1966. Checklist of the vascular plants of the Intermountain region. U. S. Forest Service Research Paper INT-32. Intermountain Forest and Range Exp. Sta. Ogden, Utah. 160 pp.

Hotchkiss, N. 1967. Underwater and floating-leaved plants of the United States and Canada. Resource Publ. 44. U. S. Dept. of the Interior. Bureau of Sport Fisheries and Wildlife. Washington, D. C.

Howell, J. T. & R. J. Long. 1970. The ferns and fern allies of the Sierra Nevada in California and Nevada. The Four Seasons 3(3):2-18.

Isely, D. 1973-. Leguminosae of the United States. Projected 3 volumes, of which 2 have been published (1. Subfamily Mimosoideae, Mem. New York Bot. Garden 25(1):1-152; II. Subfamily Caesalpinioideae, Mem. New York Bot. Garden 25(2):1-228).

Klimas, J.E. & J.A. Cunningham. 1974. Wildflowers of eastern America. A. A. Knopf.
New York. 273 pp.

Little, E. L., Jr. 1950. Southwestern trees. A guide to the native species of New Mexico and Arizona. U. S. D. A. Agric. Handbook No. 9. Washington. 109 pp.

Little, E. L., Jr. 1953. Checklist of native and naturalized trees of the United States (Including Alaska). U. S. D. A. Agric. Handbook No. 41. Washington. 472 pp.

Luer, C. A. 1975. The native orchids of the United States and Canada, excluding Florida. New York Bot. Garden. New York. 361 pp.

McMinn, H. E. & E. Maino. 1937. An illustrated manual of Pacific coast trees. Univ. of Calif. Press. Berkeley. 409 pp.

Montgomery, F. H. 1964. Weeds of Canada and the northern United States. Ryerson Press. Toronto. 226 pp.

Muenscher, W. C. 1950. Keys to woody plants. 6th ed. Comstock Publ. Co. Ithaca, N. Y. 108 pp.

Muenscher, W. C. 1955. Weeds. 2nd ed. Macmillan Co. New York. 560 pp.

New York Botanical Garden. 1905-1949; 1954-. North American flora. Series I (Vols. 1-34), Series II in progress. New York Botanical Garden. New York.

Orr, R. T. & M. C. 1974. Wildflowers of western America. A. A. Knopf. New York. 270 pp.

Peterson, R. T. & M. McKenny. 1968. Field guide to wild flowers of northeastern and north central North America. Houghton-Mifflin. Boston. 420 pp.

Petrides, G. A. 1972. A field guide to trees and shrubs. 2nd ed. Houghton-Mifflin. Boston. 428 pp.

Pohl, R. W. 1968. How to know the grasses. 2nd ed. W. C. Brown Co. Dubuque. 244 pp.

Preston, R. J. 1961. North American trees.

2nd ed. Iowa State Univ. Press. Ames. 395 pp.

Ramseur, G. S. 1960. The vascular flora of high mountain communities of the southern Appalachians. J. Elisha Mitchell Sci. 76:82-112.

Reed, C. F. 1970. Selected weeds of the United States. U. S. D. A. Agric. Handbook No. 366. 463 pp.

Rickett, H. W. 1966-1973. Wildflowers of the United States. 6 vols. (1-Northeast, 2-Southeast, 3-Texas, 4-Southwest, 5-Northwest, 6-Central Mountains and Plains). McGraw-Hill. New York.

Rosendahl, C. O. 1955. Trees and shrubs of the Upper Midwest. Univ. of Minnesota Press. Minneapolis. 411 pp.

Rydberg, P. A. 1922. Flora of the Rocky Mountains and adjacent plains. Reprinted in 1954, Hafner Publ. Co. New York. 1143 pp.

Rydberg, P. A. 1932. Flora of the prairies and plains of central North America. Reprinted (1971) by Dover Publ. Co. New York. 2 vols. 969 pp.

Seymour, F. C. 1969. Flora of New England. C. E. Tuttle. Rutland, Vermont. 596 pp.

Shetler, S. G. & H. R. Meadow (Editors). 1972. A provisional checklist of species for Flora North America. FNA Report 64. 648 pp.

Shetler, S.G. et al. 1968. Preliminary generic taxon catalog of vascular plants for Flora North America. FNA Report 8. 69 pp.

Shinners, L. H. 1962. Vegetative key to the woody Labiatae of the southeastern coastal plain. Sida 1:92-93.

Shreve, F. & I. L. Wiggins. 1964. Vegetation and flora of the Sonoran desert. Stanford Univ. Press. Stanford. 2 vols. 1740 pp.

Small, J. K. 1913. Flora of the southeastern United States. 2nd ed. Publ. by the author. New York. 1394 pp.

Small, J. K. 1933. Manual of the southeastern flora. Univ. of North Carolina Press. Chapel Hill. 1554 pp.

Small, J. K. 1938. Ferns of the southeastern United States. Reprinted in 1963 by Hafner Publ. Co. New York. 517 pp.

Smith, J. P. 1975. A key to the genera of grasses of the conterminous United States. Mad River Press. Arcata, Calif. 39 pp.

Stephens, H. A. 1973. Woody plants of the north central plains. Univ. Press of Kansas. Lawrence. 530 pp.

Stewart, A. N., L. R. Dennis, & H. M. Gilkey. 1963. Aquatic plants of the Pacific Northwest with vegetative keys. 2nd ed. Oregon State University Press. Corvallis. 261 pp.

Taylor, T. M.C. 1970. Pacific northwest ferns and their allies. Univ. of Toronto Press. Toronto. 247 pp.

Terrell, E. E. 1970. Spring flora of the Chesapeake & Ohio canal area, Washington, D. C. to Seneca, Maryland. Castanea 35:1-26.

Treshow, M., S. L. Welsh, & G. Moore. 1970. Guide to the woody plants of the mountain states. Brigham Young Univ. Press. Provo. 178 pp.

Vines, R. A. 1960. Trees, shrubs, and woody vines of the southwest. Univ. of Texas Press. Austin. 1104 pp.

Weber, W.A. 1976. Rocky Mountain flora. 5th ed. Colorado Assoc. Universities Press. Boulder. 279 pp.

Weniger, D. 1969. Cacti of the Southwest. Univ. of Texas Press. Austin. 249 pp.

Wherry, E. T. 1942. Guide to eastern ferns, 2nd ed. Univ. of Pa. Press. Philadelphia. 252 pp.

Wherry, E. T. 1961. The fern guide; northeastern and midland United States and adjacent Canada. Doubleday. Garden City, New York. 318 pp.

Wherry, E. T. 1964. The southern fern guide. Doubleday. Garden City, New York. 349 pp.

Wood, C. E., Jr. and collaborators. 1958-. Generic flora of the southeastern United States. Journal of the Arnold Arboretum.

STATE FLORAS

Alabama

Banks, D. J. 1965. A checklist of the grasses (Gramineae) of Alabama. Castanea 30:84-96.

Clar, R. C. 1972. The woody plants of Alabama. Ann. Missouri Bot. Garden 58(2):99-242.

Dean, B. et al. 1973. Wildflowers of Alabama and adjoining states. Univ. of Alabama Press.

Martin, I. R. & W. B. Devall. 1949. Forest trees of Alabama. Alabama Polytechnic Inst. Auburn. 87 pp.

Mohr, C. T. 1901. Plant life of Alabama. Contr. U. S. Nat. Herb. 6:1-921.

Alaska

Hultén, E. 1960. Flora of the Aleutian Islands and westernmost Alaska with notes on the flora of the Commander Islands. J. Cramer. Weinheim. 376 pp.

Hultén, E. 1968. Flora of Alaska and neighboring territories. Stanford Univ. Press. Stanford. 1008 pp.

Potter, L. 1962. Roadside flowers of Alaska. Roger Burt. Hanover, N. Hamp. 590 pp.

Viereck, L. A. & E. L. Little. 1972. Alaska trees and shrubs. U. S. D. A. Agric. Handbook No. 410. 265 pp.

Viereck, L. A. & E. L. Little, Jr. 1974. Guide to Alaska trees. U. S. D. A. Agric. Handbook No. 472. Washington. 98 pp.

Welsh, S. L. 1974. Anderson's flora of Alaska and adjacent parts of Canada. Brigham Young Univ. Press. Provo. 724 pp.

Wiggins, I. L. & J. H. Thomas. 1961. Flora of the Alaska slope. Univ. of Toronto Press. Toronto. 425 pp.

Arizona

Benson, L. 1969. The cacti of Arizona. Univ. of Arizona Press. Tucson. 218 pp.

Kearney, T. H. & R. H. Peebles. 1960. Arizona flora. 2nd ed. Univ. of California Press. Berkeley. 1085 pp.

McDougall, W. B. 1973. Seed plants of northern Arizona. Mus. of N. Ariz. Flagstaff. 594 pp.

Tidestrom, I. & T. Kittell. 1941. A flora of Arizona and New Mexico. Catholic Univ. Press. Washington, D. C. 897 pp.

Arkansas

Branner, J. C. & F. V. Coville. 1891. A list of the plants of Arkansas. Ann. Rep. Geol. Survey Arkansas 1888(4):155-242.

Buchholz. J. T. & E. J. Palmer. 1926. Supplement to the Catalogue of Arkansas plants. Trans. Acad. Sci. St. Louis 25:1-64.

Lang, J. M. 1969. The Labiatae of Arkansas. Proc. Ark. Acad. Sci. 20:75-84.

Moore, D. W. 1950. Trees of Arkansas. Ark. Division of Forestry and Parks. Little Rock. 119 pp.

Smith, E. B. 1969. A survey of the Arkansas Campanulaceae (including the Lobeliaceae). Proc. Ark. Acad. Sci. 23:126-130.

Smith, E. B. 1973. An annotated list of the Compositae of Arkansas. Castanea 38:79-109.

Wilcox, W. H. 1973. A survey of the vascular flora of Crittenden County, Arkansas. Castanea 38(3):286-297.

California

Baker, M. S. 1972. A partial list of seed plants of the North Coast Ranges of California, with supplements. Santa Rosa Junior College.

Benson, L. 1969. The native cacti of California. Stanford Univ. Press. Stanford. 243 pp.

Boughey, A.S. 1968. A checklist of Orange County flowering plants. Mus. of Systematic Biol., Univ. of California, Irvine. Res. Ser. No. 1. 89 pp.

Bowerman, M. 1944. The flowering plants and ferns of Mount Diablo, California. Gillick Press. Berkeley. 290 pp.

Collins, B. J. Key to coastal and chaparral flowering plants of southern California. Calif. State Univ. Foundation, Northridge. 249 pp.

Crampton, B. 1974. Grasses in California. Calif. Nat. Hist. Guides: 33. Univ. of Calif. Press. Berkeley. 178 pp.

Dawson, E. Y. 1966. Cacti of California. Calif. Nat. Hist. Guides: 18. Univ. of Calif. Press. Berkeley. 64 pp.

Ferlatte, W. J. 1974. A flora of the Trinity Alps of northern California. Univ. of Calif. Press. Berkeley. 206 pp.

Ferris, R. S. 1968. Native shrubs of the San Francisco Bay Region. California Nat. Hist. Guides: 24. Univ. of Calif. Press. Berkeley. 82 pp.

Gillett, G. W., J. T. Howell, & H.

Leschke. 1961. A flora of Lassen Volcanic National Park, California. Wasmann J. Biol. 19(1):1-185.

Grillos, S. J. 1966. Ferns and fern allies of California. Calif. Nat. Hist. Guides: 16. Univ. of Calif. Press. Berkely. 104 pp.

Holt, V. 1962. Keys for identification of wild flowers, ferns, trees, shrubs, and woody vines of northern California. Rev. Ed. National Press Publ. Palo Alto. 174 pp.

Hoover, R. F. 1970. The vascular plants of San Luis Obispo County, California. Univ. of California Press. Berkeley. 350 pp.

Howell, J. T. 1970. Marin flora: manual of the flowering plants and ferns of Marin County, California. 2nd ed. Univ. of Calif. Press. Berkeley. 295 pp.

Howitt, B. F. & J. T. Howell. 1964. The vascular plants of Monterey County, California. Univ. of San Francisco Press. San Francisco. 184 pp.

Howitt, B. F. & J. T. Howell. 1973. Supplement to the vascular plants of Monterey County, California. Pacific Grove Mus. of Nat. Hist. Assoc. 60 pp.

Jepson, W. L. 1909-1943. A flora of California. 3 vols. Incomplete. Assoc. Students Store. Univ. of Calif. Berkeley.

Jepson, W. L. 1925. A manual of the flowering plants of California. Univ. of California Press. Berkeley. 1238 pp.

Lloyd, R. & R. Mitchell. 1973. A flora of the White Mountains, California and Nevada. Univ. of Calif. Press. Berkeley. 202 pp.

Mason, H. L. 1957. A flora of the marshes of California. Univ. of Calif. Press. Berkeley. 878 pp.

McClintock, E., W. Knight, & N. Fahy. 1968. A flora of the San Bruno Mountains, San Mateo County, California. Proc. Cal. Acad. Sci., 4th Ser. 32(20):587-677.

McMinn, H. E. 1951. An illustrated manual of California shrubs. Univ. of Calif. Press. Berkeley. 663 pp.

Metcalf, W. 1968. Introduced trees of central California. Calif. Nat. Hist. Guides: 27. Univ. of Calif. Press. Berkeley. 159 pp.

Munz, P. A. 1959. A California flora. In collaboration with D. Keck. Univ. of California Press. Berkeley. 1681 pp. (In recent printings the "Supplement" is bound as a separate unit after the "Flora".)

Munz, P. A. 1961. California spring wildflowers: from the base of the Sierra Nevada and southern mountains to the sea. Univ. of California Press. Berkeley. 123 pp.

Munz, P. A. 1962. California desert wildflowers. Univ. of California Press. Berkely. 122 pp.

Munz, P. A. 1963. California mountain wildflowers. Univ. of California Press. Berkeley. 122 pp.

Munz, P. A. 1965. Shore wild flowers of California, Oregon, and Washington. Univ. of Calif. Press. Berkeley. 122 pp.

Munz, P. A. 1968. Supplement to a California flora. Univ. of Calif. Press. Berkeley. 224 pp.

Munz, P. A. 1974. A flora of southern California. Univ. of California Press. Berkeley. 1086 pp.

Niehaus, T. F. 1974. Sierra wildflowers. California Nat. Hist. Guides: 32. Univ. of Calif. Press. Berkeley. 223 pp.

Penalosa, J. 1963. A flora of the Tiburon Peninsula, Marin County, California. Wasmann J. Bio. 21(1):1-74.

Peterson, P. V. 1966. Native trees of southern California. Calif. Nat. Hist. Guides: 14. Univ. of Calif. Press. Berkeley. 136 pp.

Peterson, P. V. & P.V. Peterson, Jr. 1975. Native trees of the Sierra Nevada. Calif. Nat. Hist. Guides: 36. Univ. of Calif. Press. Berkeley. 147 pp.

Philbrick, R. N. 1972. Plants of Santa Barbara Island. Madroño 21:329-393.

Pusateri, S. J. 1963. Flora of our Sierran National Parks. Carl & Irving Printers. Tulare. 170 pp.

Raven, P. H. 1966. Native shrubs of southern California. Calif. Nat. Hist. Guides: 15. Univ. of California Press. Berkeley. 132 pp.

Raven, P. H. & H. J. Thompson. 1966. Flora of the Santa Monica Mountains, California. Student Bookstore. Univ. of California, Los Angeles. 189 pp.

Robbins, W. W., M. K. Bellue, & W. S. Ball. 1970. Weeds of California. State of Calif. Sacramento. 547 pp.

Rockwell, J. A. & S. K. Stocking. 1969. Checklist of the flora. Sequoia-Kings Canyon National Parks. Sequoia Natural History Assoc. Three Rivers, Calif. 96 pp.

Rodin, R. J. 1960. Ferns of the Sierra. Yosemite Nat. Hist. Assoc. Special issue of Yosemite Nature Notes 39(4). 79 pp.

Sharsmith, H. K. 1965. Spring wildflowers of the San Francisco Bay Region. Calif. Nat. Hist. Guides: 11. Univ. of Calif. Press. Berkeley. 192 pp.

Smith, C. F. 1976. A flora of the Santa Barbara region, California. Santa Barbara Museum of Nat. History. Santa Barbara. 331 pp.

Smith, G. L. 1973. A flora of Tahoe Basin and neighboring areas. Wasmann J. Biol. Vol. 31, No. 1. 231 pp.

Thomas, J. H. 1961. Flora of the Santa Cruz Mountains of California, a manual of vascular plants. Stanford Univ. Press. Stanford. 434 pp.

Thomas, J. H. & D. R. Parnell. 1974. Native shrubs of the Sierra Nevada. Calif. Nat. Hist. Guides: 34. Univ. of Calif. Press. 127 pp.

Thorne, R. F. 1967. A flora of Santa Catalina Island, California. Aliso 6(3):1-77.

True, G. H. 1973. The ferns and seed plants of Nevada County, California. Cal. Acad. Sci. San Francisco. 62 pp.

Twisselmann, E. C. 1956. A flora of the Temblor Range and the neighboring part of the San Joaquin Valley. Wasmann J. Bio. 14:161-300.

Twisselmann, E. C. 1967. A flora of Kern County, California. Univ. of San Francisco. San Francisco. 295 pp.

Weeden, N.F. 1975. A survival handbook to Sierra flora. Publ. by the author. 406 pp.

Witham, H. V. 1972. Ferns of San Diego County. San Diego Nat. Hist. Mus. 72 pp.

Colorado

Barrell, J. 1969. Flora of the Gunnison Basin. Natural Land Inst. Rockford, Ill. 494 pp.

Harrington, H. D. 1954. Manual of the plants of Colorado. Swallow Press. Denver. 666 pp.

Matsumara, Y. & H. D. Harrington. 1955. The ture aquatic vascular plants of Colorado. Tech. Bull. 57. Colorado Agric. Exp. Sta.

Nelson, R. A. 1970. Plants of Rocky Mountain National Park. Rocky Mountain Nature Assoc. 168 pp.

Weber, W. A. 1961. Handbook of plants of the Colorado front range. 2nd ed. Univ. of Colo. Press. Boulder. 232 pp.

Welsh, S. L. & J. A. Erdman. 1964. Annotated checklist of the plants of Mesa Verde, Colorado. Brigham Young Univ. Bull. Biol. Ser. 4(2):1-32.

Connecticut

Graves, C. B. et al. 1910. Catalogue of the flowering plants and ferns of Connecticut growing without cultivation. Bull. Conn. Geol. & Nat. Hist. Ruv. 14:1-569.

Harger, E. B. et al. 1930. Additions to the flora of Connecticut. Bull. State Geoo. & Nat. Hist. Surv. 48:1-94.

Delaware

Taber, W. S. 1937. Delaware trees. Del. State Forestry Dept. Dover. 250 pp.

Tatnall, R. R. 1946. Flora of Delaward and the eastern shore. Greenwood Bookshop. Wilmington. 313 pp.

District of Columbia

Hitchcock, A. S. & P. C. Standley. 1919. Flora of the District of Columbia and vicinity. Contr. U. S. Nat. Herb. 21. 329 pp.

Florida

Conrad, H. S. 1969. Plants of central Florida. Ridge Audubon Soc. Lake Wales. 143 pp.

Kurz, H. & R. K. Godfrey. 1962. Trees of Northern Florida. Univ. of Florida Press. Gainsville. 311 pp.

Long, R. W. & O. Lakela. 1971. A flora of tropical Florida. Univ. of Miami Press. Coral Gables. 962 pp.

Luer, C. A. 1972. The native orchids of Florida. New York Bot. Garden. New York. 293 pp.

Lakela, O. & R.W. Long. 1976. Ferns of Florida. Banyan Books. Miami. 192 pp.

Lakela, O. et al. 1976. Plants of the Tampa Bay area. Banyan Books. Miami. 198 pp.

Murrill, W. A. 1945. A guide to Florida plants. Publ. by the author. Gainsville. 89 pp.

Small, J. K. 1913. Florida trees ... Publ. by the author. New York. 107 pp.

Small, J. K. 1913. Shrubs of Forida ... Publ. by the author. New York. 140 pp.

Small, J. K. 1931. Ferns of Florida ... Science Press. New York. 237 pp.

Stern, W. L. et al. 1963. The woods and flora of the Florida keys: Capparaceae. Contr. U. S. Nat. Herb. 34(2):25-43.

Ward, D. B. 1968. Checklist of the vascular flora of Florida. Part 1. Tech. Bull. 726, Agric. Exp. Stations. Univ. of Fla. Gainsille. 72 pp.

Ward, D. B. and others. Contribution to the flora of Florida. Agric. Exp. Sta. and Univ. of Florida.

Georgia

Bishop, G. N. 1948. Native trees of Georgia. 2nd ed. Univ. Ga. School of Forestry. Athens. 96 pp.

Bostick, P. E. 1971. Vascular plants of Panola Mountain, Georgia. Castanea 36(3):194-209.

Duncan, W. H. 1941. Guide to Georgia trees. Univ. of Ga. Press. Athens. 63 pp.

Elliott, S. 1816-1824. A sketch of the botany of South - Carolina and Georgia. Fascimile published in 1971 by Hafner Publ. Co. New York. 2 vols. 743 pp.

Jones, S. B., Jr. 1974. The flora and phytogeography of the Pine Mtn. region of Georgia. Castanea 39(2):113-149.

McVaugh, R. & J. H. Pyron. 1951. Ferns of Georgia. University of Georgia Press. Athens. 195 pp.

Thorn, R. F. 1954. The vascular plants of southwestern Georgia. Am. Midl. Nat. 52:257-327.

Idaho

Bjornson, B. 1946. A key to the spring flora of southwestern Idaho. 121 pp.

Davis, R. J. 1952. Flora of Idaho. W. C. Brown Co. Dubuque. 828 pp.

St. John, H. 1963. Flora of southeastern Washington and adjacent Idaho. 3rd ed. Outdoor Pictures. Escondido, Calif. 583 pp.

Illinois

Dobbs, R. J. 1963. Flora of Henry County, Illinois. Natural Land Inst. Rockford. 350 pp.

Fell, E. W. 1955. Flora of Winnebago County, Illinois. Nature Conservancy. Washington. 207 pp.

Fuller, G. D. 1955. Forest trees of Illinois. Dept. of Conservation. Division of Forestry. Springfield. 71 pp.

Gambill, W. G. 1953. The Leguminosae of Illinois. Univ. of Ill. Press. Urbana. Ill. Biol. Monog. Vol. 22, No. 4. 117 pp.

Glassman, S. F. 1964. Grass flora of the Chicago region. Am. Midl. Nat. 72:1-49.

Jones, G. N. 1963. Flora of Illinois. 3rd ed. Univ. of Notre Dame. Notre Dame. 401 pp.

Jones, G. N. & G. D. Fuller. 1955. Vascular plants of Illinois. Univ. of Ill. Press. Urbana. 593 pp.

Mohlenbrock, R. H. 1970. The illustrated flora of Illinois. Flowering plants, flowering rush to rushes. S. Ill. Univ. Press. Carbondale, 272 pp.

Mohlenbrock, R. H. 1970. The illustrated flora of Illinois. Flowering plants, lilies to orchids. S. Ill. Univ. Press. Carbondale. 288 pp.

Mohlenbrock, R. H. 1972. The illustrated flora of Illinois. Grasses - *Bromus* to *Paspalum*. S. Ill. Univ. Press. Carbondale.

Mohlenbrock, R. H. 1975. Guide to the vascular flora of Illinois. S. Illinois Univ. Press. Carbondale & Edwardsville. 494 pp.

Mohlenbrock, R.H. 1976. The illustrated flora of Illinois. Sedges: *Cyperus* to *Scleria*. S. Ill. Univ. Press. Carbondale. 192 pp.

Mohlenbrock, R. H. & J. W. Voight. 1959. A flora of southern Illinois. S. Ill. Univ. Press. Carbondale. 390 pp.

Myers, R. M. 1972. Annotated catalog and index for the Illinois flora. West. Illinois Univ. Macomb. 64 pp.

Swink, F. A. 1974. Plants of the Chicago region. 2nd ed. Morton Arboretum. 474 pp.

Winterringer, G. S. 1967. Wild orchids of Illinois. Ill. State Mus. Popular Sci. Series. Vol. 6. Springfield.

Winterringer, G.S. & A.C. Lopinot. 1966. Aquatic plants of Illinois. Illinois State Mus. Popular Sci. Series, Vol. 6. Ill. State Mus. Div. and Dept. of Cons., Div. of Fisheries. Springfield.

Indiana

Deam, C. C. 1940. Flora of Indiana. Dept. of Conservation. Indianapolis. 1236 pp.

Peattie, D. C. 1930. Flora of the Indiana dunes: A handbook of the flowering plants and ferns of the Lake Michigan coast of Indiana and of the Calumet District. Field Mus. Nat. Hist. Chicago. 432 pp.

Iowa

Beal, E. O. 1953. Aquatic monocotyledons of Iowa. Proc. Iowa Acad. Sci. 60:89-91.

Beal, E. O. & P. H. Monson. 1954. Marsh and aquatic angiosperms of Iowa. Monocotyledons. Iowa Univ. Studies Nat. Hist. 19(5). 95 pp.

Campbell, R. B. 1961. Trees of Iowa. Agricultural and Home Economics Exp. Station, Cooperative Extension Service, Iowa State Univ. Ames. 63 pp.

Conard, H. S. 1951. Plants of Iowa, being a seventh edition of the Grinnell flora. Publ. by the author. 90 pp.

Cratty, R. I. 1933. The Iowa flora. Iowa State Coll. Journ. Sci. 7:177-252.

Crawford, D. J. 1970. The Umbelliferae of Iowa. Univ. Iowa Stud. in Nat. Hist. 21(4). 36 pp.

Davidson, R. A. 1959. The vascular flora of southeastern Iowa. State Univ. Iowa Stud. Nat. Hist. 20(2). 102 pp.

Eilers, L. J. 1971. The vascular flora of the Iowan area. Univ. Iowa Stud. Nat. Hist. 21(5). 137 pp.

Gilly, C. L. 1946. The Cyperaceae of Iowa. Iowa State College J. of Sci. 21(1):55-151.

Greene, W. 1907. Plants of Iowa. Bull. of the State Hort. Soc. Des Moines. 264 pp.

Guldner, L. F. 1960. The vascular plants of Scott and Muscatine Counties ... Davenport Publ. Mus. 228 pp.

Hayde, A. 1943. A botanical survey in the Iowa Lake Region of Clay and Palo Alto counties. Iowa State College J. of Sci. 17(3):277-416.

Pohl, R. W. 1966. The grasses of Iowa. Iowa State J. of Sci. 40(4):341-566.

Pohl, R. W. 1975. Keys to Iowa vascular plants. Kendall-Hunt. Dubuque. 198 pp.

Kansas

Barkley, T. H. 1968. A manual of the flowering plants of Kansas. Kansas State Univ. Endowment Assoc. Manhattan. 402 pp.

Gates, F. C. 1940. Annotated list of plants of Kansas: ferns and flowering plants. Kansas State Printing Plant. Topeka. 266 pp.

Stevens, W. C. 1961. Kansas wild flowers. 2nd ed. Univ. of Kansas Press. Lawrence. 461 pp.

Stephens, H. A. 1969. Trees, shrubs, and woody vines in Kansas. Univ. Press of Kansas. Lawrence. 250 pp.

Kentucky

Braun, E. L. 1943. An annotated catalog of spermatophytes of Kentucky. Publ. by the author. Cinncinati, Ohio. 161 pp.

Gunn, C. R. 1968. A check list of the vascular plants of Bullitt County, Kentucky. Castanea 33:89-106.

McFarland, F. T. 1942. A catalogue of the vascular plants of Kentucky. Castanea 7(6-7):77-108.

Meijer, W. 1972. Tree flora of Kentucky. Univ. of Kentucky. Mimeographed. 144 pp.

Murphy, G. W. 1970. A preliminary survey of the flora of Casey County, Kentucky. Castanea 35:118-131.

Spilman, C. H. 1853. Catalogue of Kentucky plants. Trans. Kent. State Med. Soc. 1852:306-318.

Wharton, M. E. 1971. A guide to the wildflowers and ferns of Kentucky. Univ. Press of Kentucky. Lexington. 344 pp.

Wharton, M. E. & R. W. Barbour. 1973. Trees and shrubs of Kentucky. Univ. Press. of Kentucky. Lexington. 582 pp.

Louisiana

Brown, C. A. 1945. Louisiana trees and shrubs. La. Forestry Commission. Baton Rouge. 262 pp.

Brown, C. A. 1972. Wildflowers of Louisiana and adjoining states. Louisiana State Univ. Press. Baton Rouge. 247 pp.

Brown, C. A. & D. A. Correll. 1942. Ferns and fern allies of Louisiana. Louisiana State Univ. Press. Baton Rouge. 186 pp.

Reese, W. D. & J. W. Thieret. 1966. Botanical study of the Five Islands of Louisiana. Castanea 31:251-277.

Robin, C. C. (Translator). 1967. Florula ludoviciana; or a flora of the State of Louisiana, by C. S. Rafinesque. Trans. from the French. Hafner. New York. 178 pp.

Maine

Ogden, E. B. 1948. The ferns of Maine. Univ. Maine. Orono. 128 pp.

Ogden, E. C., F. H. Steinmetz, & F. Hyland. 1948. Check list of the vascular plants of Maine. Bull. Joss. Bot. Soc. 8:1-70.

Wallace, J. E. 1951. The orchids of Maine. Univ. of Maine Bill. Vol. 53. Univ. of Maine Press. Orono.

Maryland

Brown, R. G. & M. L. Brown. 1972. Woody plants of Maryland. Maryland Univ. Press. College Park. 347 pp.

Kaylor, J. F. 1946. Trees of Maryland. Solomons, Dept. of Research and Education. 23 pp.

Norton, J. B. S. & R. G. Brown. 1946. A catalog of the vascular plants of Maryland. Castanea 11:1-50.

Stieber, M. T. 1971. The vascular flora of Anne Arundel County, Maryland: an annotated check list. Castanea 36(4):263-312.

Massachusetts

Eaton, R. J. 1974. A flora of Concord. Mus. of Comp. Zool. Cambridge. 236 pp.

Harris, S. K. et al. 1975. Flora of Essex County, Massachusetts. Peabody Mus. Salem.

MacKeever, F. C. 1968. Native and naturalized plants of Nantucket. Univ. of Massachusetts Press. Amherst. 132 pp.

Michigan

Billington, C. 1949. Shrubs of Michigan. 2nd ed. Cranbrook Inst. Sci. Bloomfield Hills. 339 pp.

Billington, C. 1952. Ferns of Michigan. Cranbrook Inst. Sci. Bloomfield Hills. 240 pp.

Gleason, H. A. 1939. The plants of Michigan. Wahr's Bookstore. Ann Arbor. 204 pp.

Hall, M. T. 1959. An annotated list of the plants of Oakland County, Michigan. Cranbrook Inst. of Science. Bloomfield Hills. 93 pp.

Otis, C. H. 1950. Michigan trees. 9th rev. ed. Univ. of Michigan Press. Ann Arbor. 362 pp.

Smith, H. V. 1966. Michigan wildflowers. Cranbrook Inst. of Science. Bull. No. 42 (Revised). Bloomfield Hills, Mich. 468 pp.

Voss, E. G. 1967. A vegetative key to the genera of submerged and floating aquatic vascular plants of Michigan. Mich. Bot. 6:35-50.

Voss, E. G. 1972. Michigan flora: A guide to the identification and occurrence of the native and naturalized seed - plants of the state. Part I: Gymnosperms and monocots. Cranbrook Inst. Sci. Bull. 55. 488 pp.

Minnesota

Lakela, O. 1965. Flora of northeastern Minnesota. Univ. of Minnesota Press. Minneapolis. 541 pp.

Monserud, W. & G. B. Ownbey. 1971. Common wildflowers of Minnesota. Univ. of Minnesota Press. Minneapolis. 331 pp.

Moore, J. W. & R. M. Tryon, Jr. 1946. A preliminary checklist of the flowering plants, ferns, and fern allies of Minnesota. Univ. of Minnesota Press. Minneapolis. 99 pp.

Moyle, J. B. 1964. Northern non-woody plants: a field key to the more common ferns and flowering plants of Minnesota and adjacent regions. Burgess Publ. Co. Minneapolis. 108 pp.

Mississippi

Jones, S. B., Jr. 1974. Mississippi Flora. I. Monocotyledon families with aquatic or wetland species. Gulf Research Reports 4(3):357-379.

Jones, S. B., Jr. 1974. Mississippi flora. II. Distribution and identification of the Onagraceae. Castanea 39:370-379.

Jones, S. B., Jr. 1975. Mississippi Flora. III. Distribution and identification of the Brassicaceae. Castanea 40:238-252.

Jones, S. B., Jr. 1975. Mississippi flora. IV. Dicotyledon families with aquatic or wetland species. Gulf Research Reports 5(1):7-22.

Jones, S. B., Jr. 1976. Mississippi Flora. V. The mint family. Castanea 41:41-58.

Jones, S. B., Jr., T. M. Pullen, & J. R. Watson. 1969. The pteridophytes of Mississippi. Sida 3:359-364.

Lowe, E. N. 1921. Plants of Mississippi. Bull. Mississippi State Geol. Survey 17:1-292.

Mattoon, W. R. & J. M. Beal. 1936. Forest trees of Mississippi. Mississippi State College. State College. 80 pp.

Pullen, T. M. 1966. Preliminary checklist of the Orchidaceae of Mississippi. Castanea 31:153-154.

Temple, L. C. & T. M. Pullen. 1968. A preliminary checklist of the Compositae of Mississippi. Castanea 33:106-115.

Missouri

Handebrink, E. L. 1958. The flora of southeast Missouri. Publ. by the author. Kennett, Mo. 78 pp.

Kucera, C. L. 1961. The grasses of Missouri. Univ. of Missouri Press. Columbia. 241 pp.

Settergren, C. & R. E. McDermott. 1962. Trees of Missouri. Agricultural Exp. Station, University of Missouri. 123 pp.

Steyermark, J. A. 1963. Flora of Missouri. Iowa State Univ. Press. Ames. 1725 pp.

Montana

Booth, W. E. 1950. Flora of Montana (Part I, conifers and monocots). Montana State College Research Foundation. Bozeman. 232 pp.

Booth, W. E. & J. C. Wright. 1962. Flora of Montana. Part II. Dicotyledons. Montana State College. Bozeman. 280 pp.

Drummond, J. 1949. Native trees of Montana. Mont. State Col. Ext. Serv. Bozeman. 44 pp.

Nebraska

Lommasson, R. C. 1973. Nebraska wild flowers. Univ. of Nebraska Press. Lincoln. 185 pp.

Petersen, N. F. 1923. Flora of Nebraska. 3rd ed. Lincoln. 220 pp.

Pool, R. J. 1951. Handbook of Nebraska trees. Rev. ed. Univ. Nebr. Conserv. and Surv. Div. Loncoln. 179 pp.

Winter, J. M. 1936. An analysis of the flowering plants of Nebraska. Bull. Bot. Surv. Nebraska 10:1-203.

Nevada

Archer, W. A. & Collaborators. 1940-1965. Contributions toward a flora of Nevada. Parts 1-50; incomplete. U. S. Nat. Arboretum and U. S. D. A. Washington.

Beatley, J.C. 1976. Vascular plants of the Nevada Test Site and Central-Southern Nevada: ecologic and geographic distributions. Technical Information Center, Energy Research and Development Administration. 316 pp.

Clokey, I. W. 1951. Flora of the Charleston Mountains, Clark County, Nevada. Univ. of Calif. Press. Berkeley. 274 pp.

Holmgren, A.H. 1942. A handbook of the vascular plants of northeastern Nevada. U. S. Dept. of Interior and Utah State Agric. College and Exp. Sta. Logan. 214 pp.

Lloyd, R. & R. Mitchell. 1973. A flora of the White Mountains, California and Nevada. Univ. of Calif. Press. Berkeley. 202 pp.

Tidestrom, I. 1925. Flora of Utah and Nevada. Contr. U. S. Nat. Herb. 25:1-655.

New Hampshire

Baldwin, H. I. 1974. The flora of Mount Monadnock, New Hampshire. Rhodora 76:205-228.

Foster, J. H. 1941. Trees and shrubs of New Hampshire. 2nd Ed. Soc. for Protect. of N. H. Forests. Concord. 112 pp.

Hellquist, C. B. 1971. Vascular flora of Ossipee Lake, New Hampshire and its shoreline. Rhodora 73:249-261.

Pease, A. S. 1964. A flora of northern New Hampshire. New England Bot. Club. Cambridge. 278 pp.

Scamman, E. 1947. Ferns and fern allies of New Hampshire. N. H. Acad. Sci. Durham. 96 pp.

New Jersey

Britton, N. L. 1890. Catalogue of the plants

found in New Jersey. Final Rep. Geol. Surv. New Jersey 2:1-435.

Chrysler, M. A. & J. L. Edwards. 1947. The ferns of New Jersey including the fern allies. Rutgers Univ. Press. New Brunswick. 201 pp.

Stone, W. 1973. The plants of southern New Jersey. Quarterman. Boston. pp. 25-828.

New Mexico

Tidestrom I. & T. Kittell. 1941. A flora of Arizona and New Mexico. Catholic Univ. Press. Washington, D. C. 897 pp.

Wooten, E. O. & P. C. Standley. 1915. Flora of New Mexico. Reprinted in 1971 by Hafner Publ. Co. New York. 794 pp.

New York

Gleason, H. A. 1962. Plants of the vicinity of New York. New York Bot. Gard. by Hafner Publ. Co. New York. 307 pp.

House, H. D. 1924. Annotated list of the ferns and flowering plants of New York State. New York State Mus. Bull. 254:5-759.

House, H. D. & W. P. Alexander. 1927. Flora of the Allegheny State Park region. New York State Mus. Handbook No. 2. Univ. of State of New York. Albany. 225 pp.

McVaugh, R. 1958. Flora of the Columbia County area, New York. New York State Mus. and Science Service Bull. No. 360. Univ. of New York. Albany. 400 pp.

Smith, S. J. 1965. Checklist of the grasses of New York State. Bull. New York State Mus. 403:1-44.

Zenkert, C. A. 1934. The flora of the Niagara Frontier region: Ferns and flowering plants of Buffalo, N. Y. and vicinity. Bull. Buffalo Soc. Nat. Sci. No. 16. 328 pp.

North Carolina

Blomquist, H. L. & H. J. Oosting. 1959. A guide to the spring and early summer flora of the Piedmont, North Carolina. Publ. by the authors. Durham. 181 pp.

Justice, W. S. & C. R. Bell. 1968. Wild flowers of North Carolina. Univ. of North Carolina Press. Chapel Hill. 217 pp.

Radford, A. E., H. E. Ahles, & C. R. Bell. 1965. Atlas of the vascular flora of the Carolinas. Tech. Bul. 165. North Carolina Agr. Exp. Sta. Raleigh. 208 pp.

Radford, A. E., H. E. Ahles, & C. R. Bell. 1968. Manual of the vascular flora of the Carolinas. Univ. of North Carolina Press. Chapel Hill. 1183 pp.

Tucker, G. E. 1972. The vascular flora of Bluff Mtn., Ashe County, North Carolina. Castanea 37:2-26.

Wilbur, R. L. 1963. The leguminous plants of North Carolina. N. Carolina Agric. Exp. Sta. Raleigh. 294 pp.

North Dakota

Stevens, O. A. 1963. Handbook of North Dakota plants. North Dakota Inst. for Regional Stud. Fargo. 324 pp.

Ohio

Anliot, S. E. 1973. The vascular flora of Glen Helen, Clifton Gorge, and John Bryan State Park. Ohio Biol. Sur. Biol. Notes No. 5. 162 pp.

Braun, E. L. 1961. The woody plants of Ohio. Trees, shrubs, and woody climbers native, naturalized, and escaped. Ohio State Univ. Press. Columbus. 362 pp.

Braun, E. L. 1967. The vascular flora of Ohio. Vol. 1. The Monocotyledoneae. Ohio State Univ. Press. Columbus. 464 pp.

Easterly, N. W. 1965. An illustrated guide

to the Cruciferae of Ohio. Castanea 30:177-191.

Schaffner, J. H. 1932. Revised catalog of Ohio vascular plants. Bull. Ohio Biol. Surv. 25:1-126.

Weishaupt, C. G. 1960. Vascular plants of Ohio. Publ. by the author. Columbus. 309 pp.

Wilson, H. D. 1974. Vascular plants of Holmes County, Ohio. Ohio J. Sci. 74:277-281.

Oklahoma

Goodman, G. J. 1958. Spring Flora of Central Oklahoma. Univ. of Oklahoma Duplicating Service. Norman. 126 pp.

Jeffs, R. E. & E. L. Little. 1930. A preliminary list of ferns and seed plants of Oklahoma. Univ. of Okla. Biol. Survey 2:39-101.

McCoy, D. 1976. Roadside flowers of Oklahoma. Publ. by the author. Lawton. 115 pp.

Stemen, T. R. & W. S. Myers. 1937. Oklahoma Flora. Harlow Publ. Corp. Oklahoma City. 706 pp.

Waterfall, U. T. 1969. Keys to the flora of Oklahoma. 4th ed. Publ. by the author. Dist. by the Oklahoma State Univ. Bookstore. Stillwater. 246 pp.

Oregon

Ireland, O. L. 1968. Plants of the Three Sisters region, Oregon Cascade Range. Mus. Nat. Hist. Univ. Oregon. Eugene. 130 pp.

Mason, G. 1975. Guide to the plants of the Wallowa Mountains of Northeastern Oregon. Special Publ. Museum of Natural History. University of Oregon. Eugene. 411 pp.

Peck, M. E. 1961. A manual of the higher plants of Oregon. 2nd ed. Oregon State Univ. Press. Corvallis. 936 pp.

Weidemann, A. F. 1969. Plants of the Oregon Coastal dunes. Oregon State Univ. Bookstores. Eugene.

Pennsylvania

Cannan, E. D. 1946. A key to the ferns of Pennsylvania. Publ. by the author. Johnstown. 112 pp.

Grimm, W. C. 1950. The trees of Pennsylvania. Stackpole and Heck. New York and Harrisburg. 363 pp.

Grimm, W. C. 1952. The shrubs of Pennsylvania. Stackpole Co. Harrisburg. 522 pp.

Henry, L. K. 1971. An annotated list of the vascular flora of Butler County, Pennsylvania. Ann. Carn. Mus. Pittsburgh 43(5):115-178.

Henry, L.K. & W.E. Baker. 1955. Orchids of western Pennsylvania. Ann. Carnegie Mus. 330:299-346.

Henry, L. K. et al. 1975. Western Pennsylvania orchids. Castanea 40:93-168.

Jennings, O. E. 1953. Wild flowers of western Pennsylvania and the upper Ohio Basin. Univ. of Pittsburgh Press. Pittsburgh. 2 vols.

Kelly, J. P. 1937. The ferns and flowering plants of central Pennsylvania. Penn. State Coll. State College. 120 pp.

Porter, T. C. 1903. Flora of Pennsylvania. Publ. by the author. Boston. 362 pp.

Westerfield, W. F. 1961. An annotated list of vascular plants of Centre and Huntingdon Counties, Pennsylvania. Castanea 26:1-80.

Rhode Island

Palmatier, E. A. 1952. Flora of Rhode Island. Univ. of Rhode Island. Kingston. 75 pp.

South Carolina

Batson, W. T. 1970. Wild flowers in South Carolina. Univ. of South Carolina Press. Columbia. 146 pp.

Radford, A. E., H. E. Ahles, & C. R. Bell. 1965. Atlas of the vascular flora of the Carolinas. Tech. Bull. 165. North Carolina Agr. Exp. Sta. Raleigh. 208 pp.

Radford, A. E., H. E. Ahles, & C. R. Bell. 1968. Manual of the vascular flora of the Carolinas. Univ. of North Carolina Press. Chapel Hill. 1183 pp.

South Dakota

Van Bruggen, T. 1976. Vascular plants of South Dakota. Iowa State University Press. Ames. 538 pp.

Winter, J. M. & C. K. & T. Van Bruggen. 1959. A checklist of the vascular plants of South Dakota. Dept. of Botany, State Univ. of South Dakota. Vermillion. 176 pp.

Tennessee

Anderson, W. A., Jr. 1929. The ferns of Tennessee. Univ. Tenn. Knoxville. 40 pp.

Gattinger, A. 1901. The flora of Tennessee and a philosophy of botany. Publ. by the author. Nashville. 296 pp.

Mahler, W. F. 1970. Manual of the legumes of Tennessee. J. of Tenn. Acad. of Sci. 45:65-96.

Rogers, K. E. & F. D. Bowers. 1969-. Notes on Tennessee plants. Castanea 34:294-397; 36:191-194; 38:335-339.

Shaver, J. M. 1954. Ferns of Tennessee. Geo. Peabody Coll. for Teachers. Nashville. 502 pp.

Smith, C. R. & R. W. Pearman. 1971. A survey of the pteridophytes of northeastern Tennessee. Castanea 36:181-191.

Texas

Correll, D. S. 1956. Ferns and fern allies of Texas. Texas Research Foundation. Renner. 188 pp.

Correll, D. S. & M. C. Johnston. 1970. Manual of the vascular plants of Texas. Texas Research Foundation. Renner. 1881 pp.

Gould, F. W. 1962. Texas plants - A checklist and ecological summary. A & M. College of Texas. Texas Agric. Exp. Sta. College Station. 112 pp.

Gould, F. W. 1975. The grasses of Texas. Texas A & M Univ. Press. College Station. 643 pp.

Gould, F. W. & T. W. Box. 1965. Grasses of the Texas coastal bend. Texas A & M Univ Press. College Station. 186 pp.

Johnston, E. G. 1972. Texas wild flowers. Shoal Creek Publ. Austin. 205 pp.

Jones, F. B. 1975. Flora of the Texas coastal bend. Welder Wildlife Found. Sinton. 262 pp.

Jones, F. B., C. M. Rowell, Jr., & M. C. Johnston. 1961. Flowering plants and ferns of the Texas coastal bend counties. Welder Series B-1. Welder Wildlife Foundation. Sinton. 146 pp.

Lundell, C. L. & Collaborators. 1955-. Flora of Texas. Projected 10-volume work, Vol. 1 (pt. 1) and Vol. 3 have been published. Texas Research Foundation. Renner.

Mahier, W. F. 1973. Flora of Taylor County, Texas. Southern Methodist Univ. Bookstore. Dallas. 247 pp.

Reeves, R. G. 1972. Flora of Central Texas. Prestige Press. Ft. Worth. 320 pp.

Reeves, R. G. & D. C. Bain. 1947. Flora of south central Texas. Texas A & M College. College Station. 298 pp.

Rickett, H. W. 1969. Wildflowers of the United States: Texas. Vol. 3. McGraw-Hill. New York. 553 pp.

Shinners, L. H. 1972. Spring flora of the Dallas-Fort Worth area, Texas. 2nd ed. revised by W. E. Mahler, Prestige Press. Fort Worth. 514 pp.

Turner, B. L. 1959. The legumes of Texas. Univ. of Texas Press. Austin. 284 pp.

Vines, R. A. 1953. Native east Texas trees. Publ. by the author. Houston. 131 pp.

Wills, M. M. & H. S. Irwin. 1961. Roadside flowers of Texas. Univ. of Texas. Austin. 295 pp.

Utah

Flowers, S. 1944. Ferns of Utah. Bull. Univ. of Utah. 35(7): Biol. Sci. Ser. 4(6):1-87.

Harrison, B. F., S. L. Welsh, & G. Moore. 1965. Plants of Arches National Monument. Brigham Young Univ. Sci. Bull. Biol. Ser. 5(1):1-23.

Higgins, L. C. 1972. The Boraginaceae of Utah. Brigham Young Sci. Bull. Biol. Ser. 16(3). 83 pp.

Holmgren, A. H. 1948. Handbook of the vascular plants of the northern Wasatch. Publ. by the author. 202 pp.

Tidestrom, I. 1925. Flora of Utah and Nevada. Contr. U. S. Nat. Herb. 25:1-655.

Welsh, S. L. & G. Moore. 1973. Utah plants; Tracheophyta. 3rd ed. Brigham Young Univ. Press. Provo. 474 pp.

Vermont

Seymour, F. C. 1969. The flora of Vermont. 4th edition. Agric. Exp. Sta. Bull. 660. Univ. of Vermont. Burlington. 393 pp.

Virginia

Cooperrider, T. S. & R. F. Thorne. 1964. The flora of Giles County, Virginia. II. Castanea 29:46-70.

Harvill, A. M., Jr. 1970. Spring flora of Virginia. Publ. by the author. Farmville. 240 pp.

Hathaway, W. T. & G. W. Ramsey. 1973. The flora of Pittsylvania County, Virginia. Castanea 38:38-78.

Massey, A. B. 1944. The ferns and fern allies of Virginia. Va. Polytechnic Inst. Blacksburg. 110 pp.

Massey, A. B. 1956. Legumes in Virginia. Virg. Polytech. Inst., Virginia Agric. Exp. Sta. Bull. 478. Blacksburg. 32 pp.

Massey, A. B. 1961. Virginia flora. Virginia Agric. Exp. Sta. Tech. Bull. 152. Blacksburg. 258 pp.

Mazzeo, P. M. 1972. The gymnosperms of Virginia: a contribution towards a proposed state flora. Castanea 37:179-195.

Merriman, P. R. 1930. Flora of Richmond and vicinity (exclusive of grasses, sedges and trees). Va. Acad. of Sci. Richmond. 353 pp.

Silberhorn, G. M. 1976. Tidal wetland plants of Virginia. Educational Series No. 19, Virginia Inst. of Marine Science. Gloucester Point. 86 pp.

Washington

Jones, G. N. 1938. The flowering plants and ferns of Mt. Rainier. Univ. of Washington Publ. in Biol. 7. 192 pp.

Piper, C. V. 1906. Flora of the state of Washington. Contr. U. S. Nat. Herb. 11:1-632.

St. John, H. 1963. Flora of southeastern Washington and adjacent Idaho. 3rd ed. Outdoor Pictures. Escondido, Calif. 583 pp.

West Virginia

Clarkson, R. B. 1966. The vascular flora of the Monongahela National Forest, West Virginia. Castanea 31:1-120.

Core, E. L. & N. Ammons. 1946. Woody plants of West Virginia in winter condition. Book Exchange. Morgantown. 124 pp.

Core, E. et al. 1944. West Virginia grasses. Bull. 313. Agric. Exp. Sta. Morgantown, West Virginia.

Core, E. L. et al. 1960. Plant life of West Virginia. Scholar's Library. New York. 224 pp.

Lafferty, E. R. 1973. Plants of Wyoming County, West Virginia. Castanea 38:307-322.

Strausbaugh, P.D. & E.L. Core. 1952-1964. Flora of West Virginia. 2 vols. West Virginia Univ. Morgantown. pp. 1-273; 275-570.

Strausbaugh, P. D. & E. L. Core. 1971. Flora of West Virginia: Part 2. 2nd edition. West Virginia University Bull. Series 71, No. 12:275-575.

Wisconsin

Fassett, N. C. 1939. The leguminous plants of Wisconsin. Univ. of Wisc. Press. Madison. 157 pp.

Fassett, N. C. 1951. Grasses of Wisconsin. Univ. of Wisconsin Press. 173 pp.

Fassett, N.C. 1976. Spring flora of Wisconsin. 4th ed., revised and enlarged by O.S. Thomson. Univ. of Wisconsin Press. Madison. 413 pp.

Freckmann, R. W. 1972. Grasses of central Wisconsin. Reports on the fauna and flora of Wisconsin. Report No. 6. Mus. of Nat. Hist. Univ. of Wisconsin. Stevens Point. 81 pp.

Hartley, T. G. 1966. The flora of the drift-less area. Univ. Iowa Stud. in Nat. Hist. 21(1). 174 pp.

Many authors. 1929-1963; continuing. Preliminary reports on the flora of Wisconsin. Publ. in Trans. Wisconsin Acad. Sci., Arts & Letters.

Musselman, L. J. et al. 1971. The flora of Rock County, Wisconsin. Mich. Bot. 10(4):147-193.

Tryon, R. M., Jr., and others. 1953. The ferns and fern allies of Wisconisn. 2nd ed. Univ. of Wis. Press. Madison. 158 pp.

Wyoming

Beetle, A. A. & M. May. 1971. Grasses of Wyoming. Res. J. 39. Agric. Exp. Sta. Univ. of Wyoming. Laramie. 151 pp.

Johnson, W. M. 1964. Field key to the sedges of Wyoming. Univ. Wyoming Agric. Exp. Sta. Laramie. 239 pp.

McDougall, W. B. & H. A. Baggley. 1956. Plants of Yellowstone National Park. 2nd ed. rev. Yellowstone Library and Mus. Assoc. Yellowstone Park. 186 pp.

Porter, C. L. 1944-1961. Contributions toward a flora of Wyoming. No. 1-34. Rocky Mountain Herbarium. University of Wyoming Laramie.

Porter, C. L. 1962-. A flora of Wyoming. Issued in Parts by the Agric. Exp. Sta. Univ. of Wyoming. Laramie.

Shaw, R. J. 1968. Vascular plants of Grand Teton National Park, Wyoming. Sida 4(1):1-56. Reprinted by Grand Teton Nat. Hist. Assoc. Moose, Wyo.

Shaw, R.J. 1976. Field guide to the vascular plants of Grand Teton National Park and Teton County, Wyoming. Utah State Univ. Press. Logan. 301 pp.

CULTIVATED PLANTS

Bailey, L. H. 1927. The standard cyclopedia of horticulture. 3 vols. Macmillan Co. New York. 3639 pp.

Bailey, L. H. 1949. Manual of cultivated plants. Macmillan Co. New York. 1116 pp.

Bailey, L.H. & E.Z. Bailey. 1976. Hortus Third. A concise dictionary of plants cultivated in the United States and Canada. Revised and expanded by the Staff of the Liberty Hyde Bailey Hortorium. Macmillan Publishing Co. and Collier Macmillan Publishers. New York and London. 1290 pp.

Graf, A. B. 1974. Exotica III. Pictorial encyclopedia of exotic plants. Rev. ed. Scribner. New York. 1834 pp.

Pizzetti, I. & H. Cocker. 1975. Flowers: a guide to your garden. 2 vols. Abrams, Inc. New York. 1477 pp.

Rehder, A. 1940. Manual of Cultivated trees and shrubs hardy in North America. 2nd ed. Macmillan Co. New York. 966 pp.

Wyman, D. 1971. Wyman's gardening encyclopedia. Macmillan Co. New York. 1222 pp.

USEFUL REFERENCES

Blake, S. F. 1954. Guide to popular floras of the United States and Alaska. Bibliographical Bull. No. 23. U.S. Dept. of Agriculture. Washington, D.C. 56 pp.

Blake, S. F. 1961. Geographical guide to floras of the world. Part II. Misc. Publ. 797. U. S. Dept. of Agriculture. Washington, D. C. 742 pp.

Blake, S. F. & A. C. Atwood. 1942. Geographical guide to floras of the world. Part 1. Misc. Publ. 401. U. S. Dept. of Agriculture. Washington, D. C. 336 pp.

Fish, E. 1974. Wildflowers of North America: A selected, annotated bibliography of books in print. Library of the New York Bot. Garden. Bronx. 34 pp.

Frodin, D. G. 1964. Guide to the standard floras of the world. Dept. of Botany. University of Tennessee. Knoxivlle. 59 pp.

Gunn, C. R. 1956. An annotated list of state floras. Trans. Kentucky Acad. Sci. 17(2):114-120.

Howard, A. Q. 1974. An annotated reference list to the native plants, weeds, and some of the ornamental plants of California. Cooperative Extension. University of California. Berkeley. 34 pp.

Shetler, E. R. 1965. Floras of the United States, Canada, and Greenland. Smithsonian Inst. Washington, D. C. Mimeographed. 10 pp.

8- Collection and Preparation of Vascular Plant Specimens

Practically all taxonomic research involves the use of preserved specimens. Although they often lack the aesthetic appeal of living plants, if the specimens are properly prepared they will yield vast amounts of information. They have several important uses. Specimens document the presence of a species at a particular locality. They act as a reference which aids in the identification of unknowns. The specimens are examined by experts working on revisions of species complexes, genera, etc. and those preparing local and regional floras. Because of these uses, it is very important to collect and prepare specimens in such a way that they will yield the maximum amount of information.

EQUIPMENT

The following list includes most of the items which plant collectors will find useful. While none of them is absolutely essential, having them close at hand will result in greater efficiency and higher-quality specimens.

FIELD PRESS—A series of newspapers and a few blotters between two pieces of masonite, beaverboard, heavy cardboard or similar material and bound by two ropes, belts or straps; may be small enough to fit in backpack.

PLANT PRESS—A larger version of the field press; usually 12" x 18"; consists of two end frames of wood (solid or lattice), ventilators (corrugated cardboard), blotters, all bound together by two ropes or straps.

VASCULUM—A metal or plastic box on a shoulder strap; once widely used, now largely replaced by plastic bag.

PLASTIC BAGS—Many uses.

NOTEBOOK—For field data; pocket size very useful.

DIGGING TOOL—Commonly used ones include the trowel, dandelion digger, large screwdriver, geology pick, and bricklayer's hammer.

CLIPPERS—For woody plants.

POCKET KNIFE—Useful for shrubs and trees.

TRENCHING SHOVEL—Very handy for stubborn plants.

COIN ENVELOPES—For seeds and fruits which may get separated from the plant.

PRESERVING FLUIDS—Useful for cytological and anatomical samples; alcohol, formalin, and cytological fluids

PROCEDURES IN THE FIELD

The principal purpose in plant collecting is to preserve a specimen which is as much like the living plant as possible. It is imperative to collect a complete plant whenever feasible. Do not be guilty of the dreadful practice of "top-snatching". Most herbaceous plants will yield to a digging tool; if not, try a shovel.

You must eventually identify the plants you are collecting. Proper determination will often depend upon the presence of fruits, underground structures or some other feature. These critical characters vary from one family to another. The table below summarizes the plant parts needed for identification in many standard references.

Table 24. Features emphasized in the identification of common flowering plant families.

Family	Underground Parts	Leaves	Flowers	Fruits	Take Note Of:
Amaranthaceae				X	Monoecious or dioecious
Araceae	X	X		X	
Araliaceae	X		X	X	
Asclepiadaceae			X	X	Position frting pedicel
Betulaceae				X	
Boraginaceae	X		X	X	
Cactaceae			X		Spines and glochids
Campanulaceae			X		Corolla shape
Caprifoliaceae				X	
Caryophyllaceae			X	X	Style number
Chenopodiaceae					
Commelinaceae			X		Very delicate flowers
Compositae	X	X	X	X	Ray flower color
Cornaceae				X	Branchlet color
Crassulaceae		X	X		Basal rosette
Cruciferae		X	X	X	
Cucurbitaceae				X	Monoecious or dioecious
Cyperaceae	X			X	
Ericaceae			X	X	Fruit surface
Euphorbiaceae			X	X	Watery or milky sap
Fagaceae		X		X	Growth form
Geraniaceae			X	X	

Gramineae	X	X		X	Mature Spikelets
Hydrophyllaceae	X		X	X	
Iridaceae	X		X	X	Flws. very delicate
Juglandaceae		X		X	Nature of pith
Juncaceae			X	X	Flat or terete leaves
Labiatae	X		X	X	Flw. color & markings
Leguminosae	X	X	X	X	
Liliaceae	X		X	X	Bulb morphology
Loranthaceae			X	X	Host plant
Malvaceae			X	X	Flower color
Nyctaginaceae	X			X	
Oleaceae		X		X	
Orchidaceae			X		Flw. color & markings
Plantaginaceae		X	X		
Polemoniaceae	X		X	X	Flower color
Polygonaceae	X			X	
Portulacaceae				X	
Potamogetonaceae		X		X	Stipule morphology
Ranunculaceae	X	X	X	X	
Rhamnaceae			X	X	
Rosaceae	X	X	X	X	Flwing & sterile stems
Rubiaceae				X	
Salicaceae		X		X	
Scrophulariaceae			X	X	Flw. color & markings
Solanaceae			X	X	
Ulmaceae		X		X	
Umbelliferae	X	X		X	
Violaceae		X	X		Flw. color & markings

The specimen which you collect will be glued eventually to a piece of standard-sized herbarium paper (11½" x 16½"). Keep this in mind as you collect, but do not attempt to collect only those plants which will fit comfortably within these limits. Larger plants may be folded into a "V" or "W". The folds should be obvious. A fold can be maintained by slipping a **flexostat** over the bend. Flexostats are made by cutting an index card into three or four parts. Cut a one inch slit in each and slip the folded "elbow" of the plant through the slit. After drying, the flexostats are removed and can be reused.

Larger herbaceous plants may be subdivided into a few parts, each of which will be mounted on a separate piece of herbarium paper. Most woody plants are simply too large to treat in any of these ways. The usual procedure is to remove representative parts from the plant.

Sometimes the plant or plant part is too bulky to press well. Large, often succulent parts may be sliced into thinner pieces to facilitate drying. Heavy stems can be cut to permit water to escape more readily.

Collecting aquatics can be made less painful by floating them out of the water onto wax paper or bond paper. The paper is placed beneath the plant while it is submerged in the water, and slowly raised until the specimen is out of the water and lying on the paper. Attempting to pull aquatics from the water and then arranging them on a newspaper is less than satisfactory.

Plants should be pressed as soon as possible in a single fold of newspaper. A light-weight field press can be carried directly into the field. Preliminary trimming and arranging of the plant specimens can be done on the spot. If a field press is not available, plants can be temporarily stored in tightly sealed plastic bags. A small amount of water in the bag will help to keep them fresh. Avoid exposure of the bags to direct sunlight.

Pertinent field data must accompany each specimen. Always have some sort of field notebook with you. It should contain the following information for each collection:

1) Your personal collection number. Each kind of plant collected at a particular site on a particular date should receive a different collection number. If you collect three specimens each of five different species at a particular site, you will have five collection numbers, each in triplicate. Otherwise, do not repeat numbers. If species number one is collected again at a different site, or the same site on a different day, the collection gets a new number. The first plant which you collect is "1" and the number sequence is continuous from that point. More elegant numbering schemes often lead to confusion. The number should be written on the margin of the newspaper that the plant is pressed in, so that it is permanently associated with the specimen.

2) Location. This should be as precise as possible. If topographic maps or their equivalent are available, the standard township, range, and section method of location is most useful. If not, the car odometer can provide mileage to towns, important junctions, or other stable markers. Avoid references to ephemeral places.

3) Additional information. This includes habitat type, associated species, frequency of occurrence, and notes about the plant itself, e.g. size and flower color.

4) Date. This is very helpful to the person who wants to return to your collecting site and hopes to find the plants in flower or fruit.

FINAL PRESSING AND DRYING OF SPECIMENS

After returning from the field, the specimens should be processed as quickly as possible. Field presses and plastic bags may be temporarily stored in the refrigerator, but this seldom improves the quality of the final specimen. Delay can ruin many hours of field work.

The plants should be transferred to a regular plant press. These may be purchased from scientific supply companies or you may construct your own at a fraction of the price. Pieces of wood, builder's felt, and corrugated cardboard cut to 12" x 18" will suffice. Clothesline rope or canvas straps are all that is needed to bind the press.

Final arrangement, cleaning, trimming, and folding now occur. The slightly flaccid plants are often easier to arrange at this stage than when they were collected. Flowers should be easily visible. Excess stems and leaves may be removed, but leave a stub so that it is clear that parts are missing. Remove all dirt from the root system.

There are several mothods of arranging the specimens, ventilators, and blotters to make up the press. I prefer to make a sandwich of a ventilator - blotter - specimen (in a **single** fold of newspaper) - blotter - ventilator. When pressing material is at a premium, you may wish to economize by making up repeating units of ventilator - specimen - blotter - specimen - ventilator. This will cut down on the life of the ventilators by slowly crushing the air passages. Do not build up a press more than about two feet high; it is difficult to maintain the stack when tightening the ropes.

Close the press by means of the ropes or

straps. It is good to have a substantial friend who can stand on the press. Flattening the specimens will not only decrease wrinkling of plant parts, but will also bring them into closer contact with the blotters and ventilators.

Drying may be accomplished in several ways. Perhaps the simplest is to put the press in the sunlight. Presses may also be strapped to the top of a car and dried during travel, much to the amusement of passersby. Most institutions have artificial means of drying plant specimens. These include the lightbulb drier, steam drier, or electric drier.

How long does it take to dry specimens? There is no simple answer. Too short a period will result in damp and mildewed specimens; too long in the drier will produce brittle and discolored specimens. Some plants simply take longer to dry than others. The only solution is to look at the presses often until the specimens look, feel, and smell dry. This usually takes 8-48 hours, depending upon the kinds of plants and the drier.

IDENTIFICATION

Most professional taxonomists and serious amateurs identify the plants after they have been pressed and dried. Floral and vegetative parts may be somewhat restored and made quite pliable through the use of softening agents such as **pohlstoffe.** It is composed of:

Aerosol OT*	1%
Distilled water	74%
Methanol	25%

*dioctyl sodium sulfosuccinate

The liquid is applied directly to the specimen. You may begin work in about thirty seconds. Pohlstoffe leaves no residue.

LABELS

After identification is complete, the next step is the preparation of a permanent label. Use only high-quality paper. Labels should be typed; handwritten labels are often obscure. Do not use ballpoint pens. The ink is often blurred or removed by the glue used to attach the label to the herbarium paper.

There is no standard format for label information. One possibility is shown in the photograph below. Note the following: 1) the complete scientific name (with authority) is used, 2) geographical units of decreasing size are used, 3) the dating system leaves no doubt as to the day, month, and year.

HUMBOLDT STATE COLLEGE HERBARIUM
PLANTS OF CALIFORNIA

Alisma lanceolatum With.

Humboldt Co. T6N, R1E, Sec. 18; Arcata, Siedel Road between Foster and Lamphere Road; roadside ditch; flowers light pink.

3 July 1976

4 of 4 sheets

JAMES PAYNE SMITH, JR.   8939

Completed labels should be slipped inside the newspaper with the specimen. A separate label is required for each duplicate.

MOUNTING SPECIMENS

The completed specimen and its label should remain loose in the newspaper. Never glue, tape, or staple them. Specimens should be mounted only on very high quality herbarium paper. This is generally done by the receiving institution where the specimens will be deposited. Each has its own particular way of mounting specimens. The traditional "glassplate" method involves placing the dried specimen on a sheet of glass covered with paste and then transferring it to a sheet of herbarium paper. The label is also glued to the paper, usually in the lower right hand corner. Specimens may be further secured by strapping them to the paper with thin strips of cloth tape or by sewing them.

A more recent method involves placing the specimen directly on the herbarium paper and gluing it in place with a plastic mounting medium. One common recipe for the medium is:

Dow resin 276V2	75g
Ethyl cellulose	250g
Toluene	720cc
Methanol	180cc

THE HERBARIUM

The completed specimens must be carefully stored so that they are not damaged by moisture, dirt, or insects. They must also be available for study. Both the collection of specimens and the room in which they are housed is called an herbarium. Most vascular plant herbaria are rooms filled with large metal storage cases. A few of the larger herbaria contain several million specimens. These are filed away in genus and species folders according to some system of retrieval. Older herbaria tend to use schemes developed by Engler and Prantl or Bentham and Hooker, while the newer and smaller collections are often arranged alphabetically.

USEFUL REFERENCES:

Brayshaw, T.C. 1973. Plant collecting for the amateur. Museum Methods Manual 1. British Columbia Provincial Museum. 15 pp.

DeWolf, G.P., Jr. 1968. Notes on making an herbarium. Arnoldia 28:69-111.

Radford, A.E. et al. 1974. Vascular plant systematics. Chp. 18 (pp. 387-398). Harper & Row. New York.

Smith, C.E., Jr. 1971. Preparing herbarium specimens of vascular plants. Agric. Information Bull. No. 348. U.S. Department of Agriculture. Washington, D.C. 29 pp.

Glossary

acaulescent (L., *acaulescens*, becoming stemless) stemless or apparently so; the stem often subterranean

accessory fruit- a false fruit, as in the strawberry, in which the bulk of the fleshy portion is derived from receptacle rather than gynoecium

achene (Gk., not gaping)- a dry, single-seeded, indehiscent fruit whose pericarp and seed coat are separate, except at the funiculus

acicular (L., *acus*, a needle)- needle-shaped, as in the leaves of pines

acorn (ME, *akern*, a nut)- the type of nut found in *Quercus, Lithocarpus,* etc. in which an involucre partially to ± completely encloses the fruit

actinomorphic (Gk., ray-shaped)- a flower in which the perianth parts are essentially the same size, shape, and texture within a series, and which radiate from the center of the flower

acuminate (L., *acumen*, a point)- an apex whose sides are somewhat concave and which taper to an extended point

acute (L., *acutus*, to sharpen)- an apex formed by two margins meeting at less than a 90° angle, its sides ± straight

acyclic (Gk., without + a circle)- the floral parts inserted in spirals rather than in whorls

adherent (L., *adhaero*, I stick to)- the situation in which two dissimilar structures touch, but are not actually fused

adnate (L., *adnascor*, I grow to)- the organic fusion of unlike parts, as in the union of the androecium and gynoecium

adventitious (L., *adventitius*, to + I come)- originating from mature tissues rather than meristematic ones, as in aerial roots

aerial (L., *aerius,* of the air)- growing in the air, rather than in the soil

aggregate fruit (L., *aggregatus,* assembled)- a false fruit type in which the separate carpels of an apocarpous gynoecium collectively appear to form a fruit, as in the raspberry

alternate (L., *alternatus,* to do by turns)- the leaf arrangement characterized by a single leaf per node

alveolate (L., *alveolus,* a hollow vessel)- resembling the surface of a honeycomb

ament (L., *amentum,* a strap)- see **catkin**

Amentiferae- the group name used by Engler for catkin-bearing trees and shrubs

androecium (Gk., male + house)- the floral series lying between the corolla and the gynoecium; the collective term for the stamens of a flower

anemophilous (Gk., wind + I love)- wind - pollinated

angiosperm (Gk., vessel + seed)- a semitechnical name applied to a flowering plant

annual (L., *annuus,* lasting a year)- living for a single growing season

annular (L., *annularis,* a ring)- in the form of a ring, as in the staminal disc of certain flowers and some stipular scars

anther (Gk., flowering)- the sporogenous portion or pollen-bearing part of the stamen

anthesis (Gk., flowering)- the stage during which the flower is fully expanded when pollination can occur

apetalous (Gk., without + flower leaf)- said of a flower which lacks a corolla

apex (L., *apex,* summit)- the upper or distal end of an organ; plural, **apices**

apiculate (L., *apiculum,* a little point)- a leaf apex which bears a short flexible point

apocarpous (Gk., separate + fruit)- the condition of the gynoecium in which the carpels are separate from one another

apopetalous (Gk., separate + a flower leaf)- having separate petals

aposepalous (Gk., separate + a covering)- having separate sepals

appendage (L., *appendere,* to suspend)- a subordinate part; an adjunct

appressed (L., *appressus,* kept under)- pressed against, closely applied to the surface of, as in the hairs on certain leaves

arachnoid (Gk., a spider's web)- vestiture type characterized by a cobwebby appearance

arborescent (L., *arborescens,* tree+like)- treelike, the trunk relatively short

arboreous (L., *arboreus,* tree - like)- trees, the trunk well - developed

arcuate (L., *arcuatus,* bent like a bow)- curved or bowed, as in the veins of certain leaves

areole (L., dim. of *area,* a space)- the area on the cactus stem which bears the spines and perhaps the glochids

aril (ML, *arillus,* a dried grape)- an outgrowth of the hilum or funiculus which takes the form of a partial covering around a seed

aristate (L., *arista,* an awn)- an apex which tapers to a very narrow, elongate, bristlelike point

armed (L., *arma,* implements or weapons)- equipped with sharp-pointed modified stems, leaves or epidermal outgrowths

asexual (Gk. + L., without + pertaining to sex)- not involving the union of gametes

attenuate (L., *attenuatus,* thinned)- a leaf apex or base which is characterized by a long gradual taper

auricle (L., *auricula,* ear lap)- an ear-shaped appendage, as in certain leaf bases

authority- the abbreviation of the name(s) of the person(s) who first applied the scientific name to a particular taxon

awl-shaped (ME, *awl*, a bodkin or pricker)- the leaf shape characterized by a gradual taper from the base to a sharp point, as in certain conifers

awn (ME, *awne*, sharp)- a substantial hair or bristle which terminates a plant part, as in the bracts of certain grasses

axile (L., *axila*, arm pit)- the interior angle formed by a stem and the petiole or pedicel which it bears; **axillary,** of or pertaining to an axile

axillary placentation (= **axile placentation**)- the placentation type characterized by the placentae at the center of the ovary, at the midpoint of a single septum or where several septa fuse

axis (L., *axis*, an axle)- the central stem of an inflorescence

baccate (L., *baccatus*, berried)- fleshy

banner- the uppermost petal of a caesalpinioid or papilionoid legume flower; also referred to as the **standard**

barbellate (L., *barba*, a beard)- said of hairs with barbs down the sides

basal placentation (L., *basis*, foundation)- the placentation type characterized by one to a few ovules attached at the base of the locule

basifixed (L., *basis* + *fixus*, foundation + fast)- said of a stamen whose anther is attached at its base rather than its midpoint

beak- a prominent sterile elongation of a gynoecium or fruit

beard- a line or tuft of hairs

berry- a multiseeded, indehiscent fruit in which the pericarp is fleshy throughout, as in the tomato

biennial (L., *biennum*, a period of two years)- living for two growing seasons

bifid (L., *bifidus*, cleft)- 2-parted or 2-cleft

bifurcate (L., *bifurcatus*, 2-pronged)- forked

bilabiate (L., *bilabiatus*, 2-lipped)- 2 - lipped, as in the corolla or calyx of many mints or scrophs

bipinnate (L., *bi* + *pinna*, twice +a feather)- said of a leaf which is twice pinnately compound

biseriate (L., *bi* + *serere*, twice + to join or weave together)- in two series or sets, as in a perianth consisting of sepals and petals

biserrate (L., *biserratus*, twice + a saw)- the condition in which serrations are themselves serrate; also referred to as **doubly serrate**

bisexual (L., *bi* + *sexualis*, twice + pertaining to sex)- a flower possessing both an androecium and gynoecium

biternate (L., *biternatus*, twice + in 3's)- a ternate leaf in which the first order leaflets are themselves ternately compound

blade (ME, *blad*, a leaf)- the flattened expanded portion of a leaf or petal

bract (L., *bractea*, a thin plate of metal)- a much - reduced leaf, particularly one associated with a flower or an inflorescence

bracteole- a second order bract, often associated with a subunit of a compound inflorescence

bulb (L., *bulbus*, bulb)- a subterranean plant structure consisting of a series of overlapping leaf bases inserted on a much-reduced stem axis, as in the onion

bulblet- a little bulb

caducous (L., *cadens,* falling)- falling early, especially in the sense of falling prematurely, as in the sepals of a poppy flower

caesalpiniaceous- pertaining to or resembling the flower of a caesalpinioid legume

caespitose (= **cespitose**) (L., *caespes,* a grassy field)- occurring in tufts, mats or clumps

calcarate (L. *calcar,* a spur)- spurred, as in many flowers which bear nectar spurs

calceolate (L., *calceolatus,* shoe-shaped)- slipper-shaped, as in certain orchid flowers

callosity (L., *callus,* hard skin)- a bump or raised area, as the protuberance on the calyx of certain *Rumex* species

calyx (Gk., a cup)- the outermost of the four floral series, the sepals collectively

calyx - tube - the tube of a synsepalous calyx

campanulate (L., *campana,* a bell)- bell-shaped, as in the corolla of certain blue-bells

canescent (L., *canescens,* growing gray or hoary)- vestiture type characterized by gray pubescence

capillary (L., *capillus,* a hair)- hair - like, as in the pappus of certain Compositae

capitate (L., *capitatus,* having a head)- aggregated into a dense head-like cluster

capitulum (L., *capitulum* a little head)- an inflorescence type characterized by an aggregation of ± sessile flowers on a common receptacle

capsule (L., *capsula,* a small chest)- a dry dehiscent fruit derived from a syncarpous gynoecium which opens by means of slits, lids, pores or teeth

carinate (L., *carinatus,* keeled)- having a longitudinal ridge on the dorsal or under surface, as in the keel of a boat

carpel (Gk., fruit)- the ovule - bearing structure of a flower, therefore, the basic unit of the gynoecium; generally thought to be homologous with a megasporophyll

carpophore (Gk., fruit + I carry)- an extension of the floral axis between adjacent carpels, as in the fruit of the Umbelliferae

carunculate (L., *caruncula,* a little piece of flesh)- bearing a corky bump (the **caruncle**), as on the seed of *Ricinus*

caryopsis (Gk., nut + like)- a dry, single - seeded, indehiscent fruit in which the seed and pericarp are completely fused; the fruit type of the Gramineae

catkin (L., *catulus,* reference to a cat's tail)- an inflorescence characterized by typically unisexual, apetalous, bracteate flowers in a pendant cluster; also referred to as an **ament**

caudate (L., *caudatus,* tailed)- tailed, as in the anthers of many Ericaceae and certain Compositae

caudex (L., *caudex,* stem of a tree)- the stem root axis of a plant

caulescent (L., *caulis,* a stalk)- bearing an evident aerial stem

cauline (L., *caulinus,* belonging to the stem)- pertaining to the stem; cauline leaves are attached along the stem, as opposed to being basal

cespitose-see **caespitose**

chaff (OE, *ceaf,* husk)- a reduced, thin, dry bract; used specifically for the bract subtending the flowers in many Compositae

chartaceous (L., *charta,* paper)- having a papery or tissue-like appearance; used to describe leaf and bract margins

choripetalous (Gk., separate + a flower leaf)- having separate petals; also referred to as **apopetalous** and **polypetalous**

chorisepalous (Gk., separate + a covering)- having separate sepals; also referred to as **aposepalous** and **polysepalous**

ciliate (L., *cilium*, an eyelash)- having hairs on the margins, as in certain leaves

cinereous (L., *cinereus*, ashy)- light gray

circinate (L., *circino*, I make round)- coiled upon itself lengthwise, with the apex at the center, as in unopened fern fronds

circumscissile (L., *circumscissilis*, split around)- dehiscing by means of a line which circles the fruit or anther

cladophyll (Gk., branch + leaf)- a leaflike stem, as in *Ruscus* or *Muehlenbeckia;* also referred to as a **phylloclad,** although some botanists do make a subtle distinction between the two

clambering (ME, *clamberen,* to hook oneself on)- said of vines which form a mat over other plants, typically doing so without tendrils

clasping (ME, *clapse,* to embrace)- more or less surrounding a stem, as in the lower portions of many leaves

clavate (L., *clava,* a club)- shaped ± like a baseball bat

clavuncle- the stigmatic head of the apocynaceous flower

claw (ME, *clawe,* a claw)- the narrowed, almost stalk-like base of certain petals or sepals; **clawed,** having a claw

cleft (ME, *clyft,* divided)- the condition in leaves or petals in which they are palmately or pinnately cut to about the midpoint

cleistogamous (Gk., shut + marriage)- pertaining to cleistogamy, the condition in which flowers remain closed and are self-fertilized, as in many violets

coherent (L., *coherens,* to cleave to)- the touching of similar parts or organs without their actual fusion, as in the stamens around a style

column (L., *columna,* a pillar)- the structure resulting from the fusion of the androecium and gynoecium in the orchids; also used for the tube of fused filaments in the flowers of the Malvaceae

coma (L., *coma,* the hair)- a tuft of hairs, as in those on a milkweed seed

commissure (L., *commissura,* a joint or seam)- the region where the interior faces or schizocarps or appressed stigmas and styles meet one another

comose- having a coma, a tuft of hairs

complete (L., *completus,* filled)- a flower with all four floral series

compound leaf (L., *componere,* to put together)- a leaf composed of two or more leaflets

compound pistil- a pistil composed of two or more united carpels

compressed (L., *compressus,* pressed together)- flattened

conduplicate (L., *conduplicatus,* with + folded)- folded lengthwise, the edges together, as in the leaves of many Iridaceae

connate (L., *connatus,* born at the same time)- organically united; used specifically to describe the fusion of similar structures to one another

connective (L., *connectere,* to bind together)- the region, usually interpreted as an extension of the filament, which lies between the anther-halves or thecae of an anther

connivent (L., *connivens,* winking)- see **coherent**

cordate (L., *cordatus,* heart-shaped)- heart-shaped

coriaceous (L., *corium,* leathery)- leathery

corm (L., *cormus,* a trunk)- a dense, underground vertical stem with dry papery leaf bases, as in *Gladiolus;* see also **bulb**

corniculate (L., *corniculatus,* having a little horn)- bearing a small horn-like protuberance, as in the milkweed flower

corolla (L., *corolla,* a little crown)- the floral series lying between the calyx and the androecium; composed of petals, it is typically white or brightly-colored

corolla tube- the tube of a sympetalous corolla

corona (L., *corona,* a crown)- an additional series of structures sitting between the corolla and the androecium or inserted on the corolla; often appearing as minor appendages, but sometimes quite elaborate, as in the passion flower or milkweed; variously interpreted as modifications of the corolla or androecium

corpusculum (L., *corpusculum,* a little body)- the structure which connects the translators in the milkweed flower

corymb (l., *corymbus,* a cluster of flowers)- a flat-topped or rounded inflorescence in which the pedicels are of varying length

cosmopolitan (Gk., world + a citizen)- common to all or most of the world

crenate (L., *crena,* a notch)- said of those leaf margins with shallow rounded teeth

crisped- see **undulate**

cruciform (L., *cruciformis,* cross-shaped)- cross-shaped, as in the position of the petals in the Cruciferae

culm (L., *culmus,* a stalk)- the stem of a grass plant

cuneate (L., *cuneus,* a wedge)- wedge-shaped, as in the base of certain leaves

cupule (L., *cupula,* a little cup)- the cuplike structure composed of involucral bracts subtending a flower or fruit

cuspidate (L., *cuspis,* a point)- a leaf apex which is sharply concavely constricted to a long sharp point

cyathium (Gk., a wine cup)- the reduced cymose inflorescence found in *Euphorbia* and its immediate relatives; characterized by a pistillate flower and several staminate ones with an involucre, the whole mimicing a single perfect flower

cyclic (Gk., a circle)- whorled, not spiralled

cyme (L., *cyma,* the sprout of a cabbage)- a general inflorescence type characterized by flowers which bloom from the center outward or the apex downward

cymule- a small cyme

cypsela- an achene with an adnate perianth, as in many Compositae

cystolith (Gk., cavity + stone)- a mineral deposit, usually calcium carbonate

deciduous (L., *deciduus,* ready to fall)- falling from the plant at the end of a season

decompound (L., *decompositus,* away from + put together)- more than once compound; the term most often applied beyond the twice-compound condition

decumbent (L., *decumbens,* reclining)- applied to stems which lie on the ground, but whose ends are upturned

decurrent (L., *decurrens,* running down)- leaf bases which extend downward and are adnate to a stem

decussate (L., *decussatus,* divided crosswise)- the condition in which opposite leaves at successive nodes are inserted at right angles to one another, the result being 4-ranked leaves

dehiscence (L., *dehiscens,* gaping)- the opening of a fruit or anther by means of sutures, lids, pores or teeth

deltoid (shaped like the Greek letter delta)- triangular

dentate (L., *dentatus,* toothed)- with coarse sharp teeth which are set perpendicular to the margins

denticidal capsule (L., *dens,* a tooth)- a capsule dehiscing by a series of teeth, as in the Caryophyllaceae

denticulate (L., *denticulatus,* minutely toothed)- minutely toothed

determinate (L., *determinatus,* bounded)- an inflorescence in which the terminal or central flower is the oldest, its appearance thereby terminating the cluster; the blooming sequence is often downward or outward

diadelphous (Gk., two + a brother)- the condition of an androecium in which the stamens occur in two sets, as in 5 + 5 or 9 + 1

dichasium (Gk., to disunite)- a basic unit of many cymose inflorescences, composed of an oldest central flower and two lateral ones

dichotomous (L., *dichotomus,* parted by pairs)- forked, as in the branching patterns of some stems, the venation in certain leaves or the statements in a standard key

dicot- a semitechnical group name for those flowering plants which typically have two cotyledons (hence the name), net venation, and flower parts in 4's, 5's or multiples thereof

dicotyledonous (Gk., two + a vessel)- having two cotyledons

didymous (Gk., twin)- found in pairs, as in the mericarps of the Umbelliferae or the deeply lobed silicles of certain Cruciferae

didynamous (Gk., twice + power)- the condition of having 2 stamens of one length and 2 stamens of a different length, as in many mints and scrophs

digitate (L., *digitatus,* fingered)- a compound leaf in which the leaflets arise from a single point at the end of the petiole; also referred to as **digitately compound** and **palmately compound**

dioecious (Gk., two + a house)- having staminate and pistillate flowers on separate plants of a species; the term is best applied to species, never to individual flowers

disc (= **disk**) (L., *discus,* a quoit)- an outgrowth of the receptacle or the structure resulting from the fusion of nectaries or staminodes which develops around the gynoecium; the central part of the head in the Compositae (here usually spelled **disk**)

discoid- having only disc flowers

disk flowers- the central tubular flowers of many heads in the Compositae

distichous (Gk., of two rows)- disposed in two vertical ranks, as in leaves on a stem, flowers on a rachis or florets on a rachilla

distinct (L., *distinctus,* separate)- separate within a floral series, as in the sepals separate from one another; the term **free** is not synonymous

divaricate (L., *divaricatus,* spread asunder)- spread very far apart, extremely divergent

divergent (L., *divergens,* turning)- as in divaricate, but less so

divided (L., *divisus,* parted asunder)- cut almost to the base or mid-rib, as in leaf blades or petals

dolabriform- see **malpighiaceous**

dorsifixed (L., *dorsum* + *fixus,* the back + fast)- said of anthers which are attached along their sides to the filament, rather than at their bases

doubly serrate- see **biserrate**

drupe (L., *drupa,* an olive)- a fleshy indehiscent fruit having its single seed enclosed in a stony endocarp

drupelet- a small drupe; used particularly when there is an aggregation of drupes, as in the blackberry

dull (IE, *dh(e)wel,* dark or obscure)- not shining, lacking lustre

echinate (Gk., a hedgehog)- with straight often comparatively large, prickle-like hairs

effective publication- the publication of a new scientific name, in order to be effective, must involve the distribution of printed material

elliptical (Gk., a falling short)- shaped like an ellipse, oval in general form

emarginate (L., *emarginatus,* deprived of its edge)- a leaf with a shallow notch at its apex

emergence (L., *emergo,* I emerge)- an outgrowth from a surface

endemic (Gk., a country district)- native, not introduced; confined to a given district

ensiform (L., *ensiformis,* a sword + shape)- sword-shaped

entire (ME, *enter,* untouched)- a leaf edge which lacks marginal features such as teeth, spines, etc.

entomophilous (Gk., insects + I love)- insect-pollinated

epicalyx (Gk., upon + a cup)- a set of calyx-like bracts subtending a flower, as in those of the Malvaceae

epigyny (Gk., upon + a woman)- the condition of a flower in which the ovary is below (inferior to) the points of insertion of the androecium and the perianth

epipetalous (Gk., upon + a flower - leaf)- inserted upon the ptals or corolla, applied particularly to stamens arising from these structures

epiphyte (Gk., upon + a plant)- a plant which grows upon another plant for position or support, but which does not parasitize it

equitant (L., *equitans,* riding)- folded, overlapping, 2-ranked leaves, as in many Iridaceae; also referred to as **conduplicate**

erose (L., *erosus,* gnawed)- a leaf margin which appears gnawed

even-pinnate- the condition in a compound leaf when an even number of leaflets is present, a terminal leaflet lacking

evergreen- remaining green during the dormant season

ex (L., *ex.,* from)- used to connect the name of two authors of which the second validly published a name proposed, but not validly published by the first

exfoliate (L., *exfolium,* from + a leaf)- to peel off in layers or shreds

exindusiate (L., *exindusiatus,* without + a woman's undergarment)- without an indusium

exocarp (Gk., outside + a fruit)- the outer layer of the fruit wall

exserted (L., *exsertus,* protruded)- protruding beyond, as in the stamens from the throat of the corolla

exstipulate (= **estipulate**) (L., *exstipulatus,* without + stipule)- without stipules

false fruit- the structure which results from the coalescence of the separate true fruits of an apocarpous gynoecium or an entire inflorescence

falcate (L., *falcatus*, sickle-shaped)- sickle-shaped, as in a leaf shape

farinose (L., *farina*, meal)- covered with mealiness; also referred to as **farinaceous** and **mealy**

fascicle(L., *fasciculus*, a little bundle)- a tight cluster or clump

filament (L., *filamentum*, a thread)- the stalk which supports the anther in a stamen

fibrous root system- a root system in which all of the roots are of about the same size so that none is clearly dominant, as in many monocots

filiform (L., *filiformis*, a thread + shape)- thread-like

fimbriate (L., *fimbriatus*, fringed)- fringed

floccose (L., *floccus*, a lock of wool)- a surface which is covered with tufts of soft woolly hairs, these easily removed by rubbing

floret (L., *flos*, a flower)- very small flowers, particularly those which are found in dense inflorescences, as in the Compositae; the flower, palea, and lemma of the grass spikelet

floriferous (L., *florifer*, flower-bearing)- flower-bearing

flower (L., *flos*, a flower)- the assemblage of reproductive structures and enveloping perianth; a complete flower consists of a calyx, corolla, androecium, and gynoecium

foliaceous (L., *folium* + *-aceous*, leaf + like)- leaf-like, the term applied particularly to bracts and sepals

follicle (L., *folliculus*, a small bag)- a unicarpellate, dehiscent, dry fruit which opens along one suture

foveolate (L., *fovea*, a small pit)- pitted, as in pits on a leaf surface

free (ME, not in bondage)- not organically fused to another organ, as in the stamens separate from the petals

free central placentation- the placentation type characterized by numerous ovules attached to a central, columnar, compound placenta which is inserted at the base of the locule, but which is free at its apex

frond (L., *frons*, a leaf)- the leaf of a fern or of a palm

fruit (L., *fructus*, fruit or produce)- a ripened ovary along with any adnate structures which mature along with it

fruticose (L., *fruticosus*, shrubby)- shrub-like with more than one major stem

fugacious (L., *fugax*, fleeting)- perishing quickly; falling or withering early

funiculus (L., *funis*, a rope)- the stalk which connects the ovule to the placenta

funnelform (L., *infundibulum*, a funnel)- the shape of an actinomorphic sympetalous corolla whose tube gradually widens upward; also referred to as **infundibuliform**

furrowed (L., *porca*, a furrow)- with longitudinal grooves or channels

fusiform (L., *fusiformis*, spindle-shaped)- narrow on both ends and swollen in the middle

galea (L., *galea*, a helmet)- a petal that is hollow and vaulted, thereby seeming to form a helmet over other floral parts, as in the monk's hood and certain mints; **galeate,** having a galea

gamopetalous (Gk., marriage + a flower-leaf)- a corolla in which the petals are united, thereby forming a single structure; also referred to as **sympetalous**

gamosepalous (Gk., marriage + a covering)- a calyx in which the sepals are united, thereby forming a single structure; also referred to as **synsepalous**

geniculate (L., *geniculatus*, with bent knees)- bent sharply, as at a knee

genus (L. *genus*, a race)- a taxonomic group of related species; the first element of a scientific name

gibbosity (L., *gibber*, hump-backed)- a pouch-like enlargement on one side toward the base of an organ such as the calyx or corolla; **gibbous**, having a gibbosity

glabrous (L., *glaber*, without hairs)- without hairs

gland (L., *glans*, an acorn)- a secretory structure; used more broadly for any warty protuberance; also refers to the corpusculum in the milkweed flower; **glandular**, having glands

glaucous (Gk., bluish gray)- the term is used to denote a blue-gray or sea-green color and also for a whitish waxy covering which is easily rubbed off

globose (L., *globus*, a sphere)- nearly spherical

glochid (Gk., an angular end or barb)- an apically barbed hair or bristle; **glochidiate**, having glochids

glume (L., *gluma*, husk of corn)- one of the two sterile bracts at the base of a typical grass spikelet

glutinous (L., *glutinosus*, sticky)-covered with a sticky exudate

grain- see **caryopsis**

gynandrium (Gk., female + a stamen)-the combined androecium and gynoecium of the orchid flower; also referred to as a **column** or **gynostemium**

gynoecium (Gk., female + a house)- the innermost of the four floral series; the collective term for the carpels; the female portion of the flower

gynophore (Gk., female + I carry)- the stalk within a flower upon which the gynoecium may sit

gynostegium (Gk.,female + a roof)- the structure formed by the union of the androecium and gynoecium in the Asclepiadaceae

gynobasic style (L., *gynobasis*, female + a pedestal)- a style which is inserted on the receptacle, in addition to the base of the carpels, as in the deeply 4-lobed ovary of many mints

habit (L., *habitus*, appearance)- the general appearance of a plant

hood (ME, *hod*, hat)- a component of the corona in the milkweed flower

half-inferior- the situation in which the hypanthium or receptacle is adnate to the gynoecium, but extends only about half-way up the ovary

halophyte (Gk., salt + a plant)- a plant which grows in soils of high salt content

hastate(L., *hasta*,a spear)- having the general shape of an arrowhead, but with the basal lobes turned outward ± at right angles

head (ME *hede*, head)- a short condensed spike; also referred to as a **capitulum**

helicoid cyme (Gk., twisted + cyme)- a cymose inflorescence in which the lateral branches develop only on one side, thereby producing a coiled flower cluster

herb (L., *herba*, plant)- an annual or perennial plant which dies back to the ground at the end of the growing season because it lacks the firmness resulting from secondary growth; **herbaceous**, having the features of an herb

hermaphroditic (Gk., the son of Hermes and

Aphrodite)- bisexual; a flower having both stamens and carpels

hesperidium (after the fruit of the Hesperides)- a fleshy indehiscent fruit with conspicuous septations lined with succulent hairs; the fruit of the citrus group

hip (ME, *hepe*, a briar)- vase-like leathery hypanthium containing several seed-like achenes; the false fruit of the rose

hirsute (L., *hirsutus*, rough)- with coarse more or less erect hairs

hirtellous- minutely hirsute

hispid (L., *hispidus*, bristly)- with long, rigid, bristly hairs

hispidulous- minutely hispid

hoary (OE, *har*, venerable)- covered with a white to gray-white pubescence

homochlamydeous (Gk., one in the same + mantle)- a perianth which is not distinguished into sepals and petals; **tepal** may be used to designate a perianth segment in this situation

homosporous (Gk., one in the same + seed)- producing spores of one kind only, as in most ferns

hypanthium (Gk., beneath + a flower)- a shallow cup-like to elongate tubular structure, resulting from the fusion of the perianth and the androecium, which surrounds the gynoecium; in some instances the hypanthium appears to be formed from receptacle tissue

hypogynous (Gk., under + a female)- the condition of the flower in which the insertion of the perianth and androecium is below that of the gynoecium

imbricate (L., *imbricatus*, covered with gutter tiles)- overlapping, as in cone scales or leaves on a stem

imperfect (L., *imperfectus*, not complete)- unisexual; a flower which bears either stamens or carpels, but not both

in (L., *in*, in)- used to connect the names of two authors when the first supplied a name with a description in a work actually published by the second author

incised (L., *incisus*, cut into)- cut deeply, sharply, and often irregularly into a leaf or petal margin

included (L., *inclusus*, shut in)- not protruding from a structure, as in the stamens not projecting from a corolla

incomplete(L., *incompletus*, not finished)- lacking any of the four floral series

indehiscent (L., *indehiscens*, not + gaping)- said of a fruit or anther which does not open by means of sutures, lids, poes or teeth

indeterminate (L., *indeterminatus*, not bounded)- an inflorescence in which the basal or outer flower is the oldest and the blooming sequence is upward or inward; the appearance of the first flower does not, therefore, determine the size of the inflorescence

indumentum (L., *indumentum*, a garment)- used broadly for any covering of an organ; used by some authors for a particularly dense hairiness

indurate (L., *induro*, I harden)- hardened, as in seed coats or fruit walls

indusium (L., *indusium*, a woman's undergarment)- an epidermal outgrowth of the fern leaf which covers or contains the sporangia

inferior (L., *inferior*, lower)- below, lower than, beneath; an inferior ovary is one in which the perianth and androecium are inserted above the ovary

inflorescence (L., *infloresco*, I begin to bloom)- the arrangement of flowers on a floral axis; also used to designate the sequence of flowering within a flower cluster

insectivorous (L., *insectum* + *voro*, an insect + I devour)- insect-catching, as in the sundew or bladderwort

inserted (L., *insertus*, put into)- joined to or placed on, as in the stamens joined to the receptacle

internode (L., *internodium*, between + the knot)- the portion of the stem between two successive nodes

involucel (L., *involucrum*, a wrapper)- a series of bracts which subtend a subunit of an inflorescence, as in the umbellet of a compound umbel

involucre (L., *involucrum*, a wrapper)- a series of bracts which subtend a flower or inflorescence; **involucrate**, having an involucre

investiture- see **vestiture**

irregular flower (L., *irregularis*, not according to rule)- a flower which cannot be divided into two equal halves, as in *Canna;* often used as synonymous with zygomorphic, but used here in the more restricted sense

jointed (L., *junctus*, joined)- having points of separation or what appears as such points

keel (ME, *kele*, a ship)- the two anterior united petals of a papilionoid legume flower; a raised or pronounced ridge; **keeled,** having a keel

labellum (L., *labellum*, a little lip)- a lip, applied particularly to the enlarged and often elaborate lower petal of the orchid flower

labiate (L., *labiatus*, lipped)- having a lip or lips, as in many corollas; also used to designate members of the Labiatae

lacerate (L., *lacer*, mangled)- torn or irregularly cleft, as in a leaf margin

laciniate (L., *lacinia*,the flap of a garment)- a slash or slender pointed lobe of a leaf or petal

lamina (L., *lamina*, a thin leaf)- the expanded part of a leaf; the blade

lanate (L., *lanatus*, woolly)- covered with woolly intergrown hairs

lanceolate (L., *lanceolatus*, armed with a little lance)- narrow, tapering on both ends and widest above the middle, as in a leaf shape

lanulose- short lanate

latex (L., *latex*, juice)- a white, yellowish or reddish thickened colloidal sap, as in the spurges and milkweeds

leaf (ME, *lefe*, to peel off)- the principal flattened lateral organ borne on a stem

leaflet- any of the divisions of a compound leaf

leaf gap- the break in the continuity of vascular tissue in the stem which results from the exit of a leaf trace

leaf trace- the branch of vascular tissue from the stem which extends into a leaf

legume (L., *legumen*, a pulse)- a unicarpellate dry fruit which dehisces along both sutures; the fruit type of the Leguminosae and one of several common names used for the family

lemma (Gk., a husk)- one of two bracts of the grass floret, the other being the palea, which subtends the flower

lenticular (L., *lenticularis*, lentil-shaped)-

shaped like a doubly-convex lens, as in certain seeds and fruits

lepidote (Gk., scaly)- covered with small bran-like scales

liana (Sp., *liar*, to tie)- a woody climber with elongate, flexible, non-self-supporting stems

ligulate (L., *ligulatus*, having a little tongue)- shaped like a strap or narrow band, as in a petal or the corolla of ray flowers in the Compositae

ligule (L., *ligula*, a little tongue)- a strap-shaped organ; also the membranous projection at the junction of the sheath and lamina in grasses

limb (L., *limbus*, a border or hem)- the expanded portion or border of a sympetalous corolla

linear (L., *linea*, a thread or line)- several times longer than wide; used particularly to describe a leaf shape, as in the grasses

lip (OE, *lippa*, lip)- either the upper or lower portions of a bilabiate calyx or corolla; see also **labellum**

lobe (Gk., the lower part of the ear)- any part of an organ, particularly any rounded division, as in a corolla lobe

lobed- indented about ¼ to almost ½ way to the midrib or base of a leaf blade or petal

locule (L., *locus*, a small place)- a cavity or chamber within an ovary, anther or fruit

loculicidal capsule- a capsule which dehisces by means of openings into the locules, about midway between the partitions

lodicule (L., *lodicula*, a small coverlet)- one of the two or three minute hygroscopic scales of the grass flower; generally thought to be reduced calyx segments

loment (L., *lomentum*, bean-meal)- a legume with pronounced constrictions between the seeds; usually dehiscing transversely, rather than longitudinally as in most legumes

lustrous- shining

lyrate (Gk., a lute or lyre)- a pinnatifid leaf with an enlarged terminal lobe and smaller lateral ones, as in the dandelion

macrophyll (Gk., long + a leaf)- a leaf whose vasculature is complex enough to produce a gap in the stele; also referred to as a **megaphyll**

malpighiaceous (relating to the Malpighiaceae)- with forked hairs attached at the middle, as in a miner's pick

marescent L., *marescens*, withering)- withering, but the structure remaining on the plant rather than falling

marginal placentation (L., *margo*, the edge or boundary of a body)- the placentation type in which the ovules are attached along a single suture on the ovary wall; found only in a unicarpellate or apocarpous gynoecium

mealy- with swollen hairs which collectively form a covering resembling cooking meal

megaphyll- see **macrophyll**

megaspore (Gk., large + a seed)- the larger of the two spores produced by a heterosporous plant

megasporophyll (Gk., large + seed + leaf)- a fertile leaf which bears megaspores; a carpel is generally thought to be homologous with a megasporophyll

mericarp (Gk., a part + fruit)- a portion of a fruit which splits away and functions in itself as a fruit, as in a segment of a schizocarp or the nutlets of a mint

merous (L., a suffix denoting the number of parts)- referring to the number of parts which constitute a structure, as in a 3-merous flower

mesocarp (Gk., in the middle + a fruit)- the middle layer of a fruit wall

microphyll(Gk., small + a leaf)- a leaf whose vasculature is not sufficient to produce a gap in the stele

microspore(Gk., small + a seed)- the smaller of two kinds of spores produced by heterosporous plants

microsporophyll (Gk., small + seed + leaf)- a fertile leaf which bears microspores; the stamen in the flowering plants is thought to be homologous with a microsporophyll

midrib- the principal nerve of a leaf

minute (L., *minutus,* small)- small, as in the size of your vocabulary if you had to consult a glossary for this term

monadelphous(Gk., one + brother)- the condition of the androecium when the stamens are united by their filaments into a single group

monocot- a semitechnical group name for those flowering plants which have a single cotyledon (hence the name), parallel venation, and flower parts in 3's or multiples thereof

monocotyledonous (Gk., one + a vessel)- having one cotyledon

monoecious (Gk., one + a house)- having staminate and pistillate flowers on the same plant; note that the term is not synonymous with **imperfect** and should never be applied to individual flowers

mucro (L., *mucro,* a sharp point)- a sharp terminal point or spiny tip, as in the apex of a leaf or bract; **mucronate,** having a mucro

multiple fruit- a type of false fruit in which several true fruits from separate flowers coalesce to produce a single structure which resembles a fruit, as in the Osage orange or pineapple

muricate (L., *muricatus,* pointed)- said of a surface characterized by short, hard, tubercular outgrowth

naked (AS, *nacod,* naked)- when applied to a flower, meaning without a perianth; when applied to a captiulum of the Compositae, meaning without chaff; when applied to a sorus, meaning without an indusium

nectary (Gk., drink of the Gods)- a nectar-secreting structure, often appearing as a glandular projection or as a pit

net venation (IE, *ned,* to trust together)- a venation pattern characterized by major and minor veins which form a complex network or reticulum; also referred to as **netted**

node(L., *nodus,* a knot)- the point or region on a stem where one or more leaves are borne

nut(L., *nux,* nut)- a dry, hard, indehiscent, 1-seeded fruit derived from a syncarpous gynoecium

nutlet- a small nut

obcompressed (L., *obcompressus,* inverse + pressed together)- flattened from the front and back, rather than from the sides, as in the achenes of certain Compositae that are flattened at a right angle to the radius of the common receptacle

obcordate (L., *obcordatus,* inverse + heart - shaped)- of the shape of an inverted heart

oblanceolate (L., *oblanceolatus,* inverse + armed with a little spear)- of an inverted lanceolate shape

oblique (L., *obliquus*, slanting)- slanting or unequal-sided, as in an elm leaf

oblong (L., *oblongus*, rather long)- much longer than broad, the sides being ± parallel

obovate (L., *obovalis*, inverse + egg)- of the shape of an inverted egg

obsolescent (L., *obsolescens*, worn out)- nearly obsolete; an organ which is so rudimentary that it is practically missing

obtuse (L., *obtusus*, blunt)- an apex formed by two lines which meet at more than a right angle

ocrea (= **ochrea**) (L., *ocrea*, a piece of armor which covers the leg from the knee to the ankle)- a tubular nodal sheath, formed by the fusion of two stipules; used particularly for this structure in the Polygonaceae

odd-pinnate- a pinnate leaf with a terminal leaflet, the number of leaflets being odd

opposite (L., *oppositus*, standing in front)- the leaf arrangement characterized by two and only two leaves at a node, these on opposite sides of the stem

orbicular (L., *orbicularis*, circular)- having a flat body with a circular outline, as in the shape of the leaf blade in *Hydrocotyle*

ours- used in family descriptions in this text to refer to the species that occur in the U.S., as opposed to what the family might be like in other parts of its range

ovary (L., *ovum*, an egg)- the lower swollen portion of the gynoecium which contains the ovules

ovate (L., *ovatus*, egg-shaped)- of the shape of a longitudinal section through a hen's egg; egg-shaped

ovulate (L., *ovulatus*, having eggs)- possessing ovules, as in certain conifer strobili

ovule (L., *ovum*, egg)- the structure within the ovary which, after fertilization, will become the seed; a young seed

ovuliferous (L., *ovulifero*, egg + I carry)- bearing ovules, as in the scales of gymnosperm cones

palate (L., *palatum*, the palate)- the rounded projection in certain zygomorphic corollas which ± blocks off or closes the throat of the flower

palea (L., *palea*, chaff)- the bract which immediately subtends the flower in a grass spikelet; along with a lemma and flower, the palea is a component of the floret; also the scale-like calyx remnants in certain Compositae

palmate (L., *palmatus*, having the palm of the hand)- lobed, cleft, parted, divided or compounded so that the sinuses or leaflets point to the apex of the petiole

pandurate (L., *panduratus*, having a pandura- a musical instrument)- fiddle-shaped; obovate with a pair of well - developed basal lobes

panicle (L., *panicula*, a tuft)- a much-branched raceme in which the flowers are borne on the ultimate branchlets

paniculate (L., *paniculatus*, having a tuft)- having or resembling a panicle

pantropical (Gk., all + of the solstice)- extending through tropical areas of the Old World and New World

papilionaceous (L., *papilio* + *-aceous*, butterfly + like)- of the shape of the flower or corolla of the papilionoid legume, with its prominent banner or standard, paired wing petals, and paired keel petals which are typically fused to form a keel

papillate (L., *papillatus*, bud-shaped)- having small pimple- or nipple-like protuberances

pappus (L., *pappus*, plant-down)- the modified calyx of the flower in the Compositae; usually takes the form of capillary bristles, plumose bristles, scales, awns or a low crown

parallel (Gk., parallel)- extending in the same direction and equidistant, as in the vein pattern in most monocot leaves

parietal (L., *parietalis*, belonging to a house wall)- the placentation type of a syncarpous gynoecium in which the ovules are attached at two or more points on the ovary wall or on intrusions of the wall into the locule, as in the Cucurbitaceae; some authors include marginal placentation in this type

parted (L., *partitus*, a partition)- cut almost to the base or midrib, as in a leaf or petal

pectinate (L., *pectinatus*, like a comb)- said of parts which are arranged like the teeth of a comb

pedicel (L., *pediculus*, a little foot)- the stalk which supports a single flower; **pedicellate,** having a pedicel

peduncle (L., *pedunculus*, a little foot)- the stalk which supports an inflorescence including that stalk which supports a solitary flower

peltate (L., *pelta*, a small shield)- attached to a supporting stalk at a point inside the margin, as in the petiole of a leaf or the stalk of certain cone scales

pendulous (L., *pendulus*, hanging)- hanging down, as in inflorescences or fruits

pepo (L., *pepo*, the pumpkin)- a berry with a leathery rind; derived from an inferior ovary; use often restricted to the fruit of the Cucurbitaceae

perennial (L., *perennis*, lasting through the years)- a plant which lives for three or more years, often flowering and fruiting repeatedly

perfect (L., *perfectus*, complete)- bisexual; having both stamens and carpels in the same flower

perfoliate (L., *perfoliatus*, through a leaf)- a sessile leaf or bract whose base surrounds the stem so that it appears that the stem penetrates the leaf or bract

perianth (Gk., about + a flower)- the calyx and corolla collectively

pericarp (Gk., about + a fruit)- the fruit wall, often differentiated into an exocarp, mesocarp, and endocarp

perigynium (Gk., about + a female)- the membranous sac or sheath which surrounds the gynoecium and later the achene in *Carex*

perigynous (Gk., about + a female)- a flower whose gynoecium sits within an open cup - like to tubular hypanthium, the distinct portions of the sepals, petals, and stamens arising from its edge

perisperm (Gk., about + a seed)- storage material of nucellar origin in the seed

persistent (L., *persistens*, persevering)- remaining attached

personate (L., *personatus*, masked)- a corolla with a palate which closes off the throat

petal (Gk. a flower-leaf)- a component part of the corolla

petaloid- resembling a petal, as in certain brightly-colored sepals

petiole (L., *petiolus*, a little foot or leg)- the leaf stalk; the portion of a leaf which supports the blade

petiolule- the stalk of a leaflet

phrase name- the descriptive phrases used to designate plants before binomial nomenclature was accepted; also referred to as **polynomials**

phyllary (Gk., a leaf)- a bract in an involucre; applied particularly to the bracts surrounding the common receptacle in the Compositae

phylloclad- see **cladophyll**

phyllode (= **phyllodium**)-a leaf-like petiole, as in certain *Acacia*

phyllotaxy (Gk., a leaf + arrangement)- the arrangement of leaves on stems; in its simplest terms expressed as alternate, opposite or whorled

pilose (L., *pilosus,* hairy)- covered with soft distinct hairs

pinna (L., *pinna,* a feather)- a leaflet of a pinnately compound leaf; applied more commonly to fern leaves than flowering plants; plural, **pinnae**

pinnate- with leaflets arranged on both sides of a common axis

pinnatifid (L., *pinnatifidus,* a feather + cut)-cleft, parted, or divided in a pinnate fashion

pinnule- a secondary pinna or leaflet of a leaf which is at least twice-pinnately compound

pistil (L., *pistillum,* a pestle)- a component of the gynoecium, consisting of a single stigma, style, and placenta; see also **simple pistil** and **compound pistil**

pistillate (L., *pistillatus,* having a pestle)- a flower which contains a functional gynoecium, but which lacks well-developed functional stamens; a female flower

pitted (L., *puteus,* a well)- having small cavities or depressions; also referred to as **punctate**

placenta (L., *placenta,* a cake)- the point or region where ovules are attached by their funiculi to the ovary wall; not a distinct tissue or organ, hence not comparable to the mammalian placenta; plural, **placentae**

placentation- the arrangement of placentae within the ovary

plaited (L., *plicare,* to fold)- a flattened fold, as of a petal folded back on itself; pleat

plicate- folded into plaits

plumose (L., *plume,* the down of a feather)-feathered, feather - like; used in reference to hairs or bristles to indicate very fine side branches arising from the central shaft

pholstoffe (Pohl + **stoffe**, substance)-a distilled water, methanol, and Aerosol-OT mixture used to soften dried plant specimens to facilitate examination of them; named after Richard W. Pohl, eminent and eccentric American agrostologist; also referred to as **eau de Pohl**

pollinium (L., *pollen, a fine flour)-* a body composed of the pollen grains of an anther sac; a coherent waxy mass of pollen, as in the Orchidaceae and Asclepiadaceae; plural, **pollinia**

polycotyledonous (Gk., many + a hollow)- having several cotyledons, as in many gymnosperms and a few flowering plants

polygamodioecious (Gk., many + marriage + two + a house)- said of a dioecious species which has some perfect flowers

polygamomonoecious (Gk., many + marriage+ one + a house)- said of a monoecious species which has some perfect flowers

polygamous (Gk., many + marriage)- bearing both unisexual and bisexual flowers on the same plant

polynomial- see **phrase name**

polypetalous (Gk., many + a flower-leaf)- having a corolla of separate petals; please note that the etymology is misleading, in that the term has nothing to do with the number of petals; also referred to as **apopetalous**

polysepalous (Gk., many + a covering)- having a calyx of separate sepals; please note that the etymology is misleading, in that the term has nothing to do with the number of sepals; also referred to as **aposepalous**

pome (L., *pomum,* a fruit)- a fleshy indehiscent fruit derived from an inferior ovary surrounded by an adnate hypanthium, as in the apple, pear, and related genera of the Rosaceae

poricidal (Gk., + L., a passage + to cut)- said of a capsule or anther which open by means of a pore or series of pores

prickle (OE, puncture)- a sharp-pointed epidermal outgrowth, irregularly spaced and without vasculature, as in structures on a rose stem; the term is not synonymous with **spine** or **thorn**

prostrate (L., *prostratus,* thrown to the ground)- lying flat on the ground; also referred to as **procumbent,** particularly if not rooting at the nodes

pruinose (L., *pruina,* hoar-frost)- having a waxy powdery secretion ("bloom") on the surface

pseudanthium (Gk., false + flower)- an inflorescence which mimics a single flower, as in the Compositae

puberulent (L., *puber,* downy)- downy; the hairs soft, straight, and erect; minutely pubescent

pubescence (L., *pubescens,* reaching puberty or hairiness)- a general term for hairiness

pubescent- covered with soft down hairs

pulvinus (L., *pulvinus,* a cushion)- a swollen area at the base of a leaflet or leaf which is sensitive to movement and heat

punctate (L., *punctum,* a point)- a surface characterized by dots, pits or depressions

pyrene (Gk., a kernel or stone)- the stone or pit of a drupe; the seed and endocarp of the "stone fruits"

pyxis (L., *pyxis,* a box)- a capsule which dehisces by means of a lid

raceme (L., *racemus,* a bunch of grapes)- an elongate, unbranched, indeterminate inflorescence of pedicellate flowers

racemose- applied to those inflorescence types in which the blooming sequence is from the base upward or from the outside inward; the appearance of the first flower of the inflorescence does not mark its upper limit, the blooming sequence being, therefore, indeterminate; also used to indicate an inflorescence which is raceme-like in its general appearance

rachilla (Gk., dim. of backbone)- the internal axis of the spikelet in the grasses and sedges

rachis (Gk., backbone)- the axis of an inflorescence or of a pinnately compound leaf

radial symmetry (L., *radius,* the spoke of a wheel)- the type of symmetry displayed by a flower whose perianth parts radiate from its center, such that any of many planes passed through the flower would yield two mirror images

rank (OE, *ranc,* tall and slender)- a vertical row; usually designated by a numerical prefix, as in 3-ranked leaves, meaning that the leaves appear in three vertical rows when viewed from above; please note that the determination of rank has nothing to do with the number of leaves at a given node

ray (L., *radius*, the spoke of a wheel)- the outer or ligulate flowers of a composite head; the stalk which supports an unbellet of a compound umbel

receptacle (L., *receptaculum*, a reservoir)- the ± expanded apex of the pedicel upon which the flower series are inserted

recurved (L., *recurvus*, bent backward)- curved backward or downward

reflexed (L., *reflexus*, bent back)- abruptly bent or turned downward or backward

regular flower (L., *regularis*, according to rule)- a flower which displays radial symmetry

reniform (L., *reniformis*, kidney + shaped)- kidney-shaped

repand (L., *repandus*, bent backward)- with an uneven margin, but not so pronounced as in sinuate

repent (L., *repens*, creeping)- creeping or sprawling plants, often rooting at the nodes; also referred to as **trailing**

replum (L., *replum*, a door-case)- the central papery partition in a silicle or silique away from which the fruit wall falls

resupinate (L., *resupinatus*, bent back)- upside down; twisted 180⁰, as in the orchid flower, such that the upper petal is rotated to the lowermost position

reticulate (L., *reticulatus*, netted)- netted, as in the venation pattern of many dicot leaves

retuse (L., *retusus*, blunted)- with a shallow notch at a rounded apex

revolute (L., *revolutus*, rolled back)- rolled back from the margin or apex toward the lower side

rhizoid (L., *rhizoideus*, root + like)- a root - like structure in certain vascular plants which carries on the functions of a root, but which lacks its anatomical complexity

rhizomatous- having rhizomes

rhizome (L., *rhizoma*, a root)- an underground horizontal stem which bears reduced scaly leaves

rib (ME, *ribbe*, to arch over)- a primary vein, as in the midrib of a leaf or bract

rootstock- a subterranean stem; used particularly to designate a rhizome

rosette (dim. of F., *rose*, a rose)- a cluster of leaves in a radiating pattern, usually at the base of the plant

rosulate (L., *rosulatus*, like a small rose)- forming a rosette

rostellum (L., *rostellum*, a small beak)- a small beak, applied particularly to the narrow extension of the upper edge of the stigma in certain orchids

rotate (L., *rotatus*, wheel-shaped)- a sympetalous actinomorphic corolla with a very short tube and a flat circular limb; said of a flower which is ± dish-shaped

rounded (L., *rotundus*, a wheel)- said of an apex which is gently curved

rugose (L., *rugosus*, wrinkled)- covered with wrinkles

runcinate (L., *runcinatus*, having a large saw)- coarsely toothed to incised, the teeth pointed toward the base of the leaf

runner- see **stolon**

saccate (L., *saccatus*, bag-shaped)- shaped like a pouch or bag, as in the lower petal of certain orchids

sagittate (L., *sagittatus*, arrow-shaped)- shaped generally like an arrow-head, with the basal lobes pointed downward; see also **hastate**

salverform (Sp., *salva,* foretasting; hence a tray + shaped)- a sympetalous actinomorphic corolla with an elongate slender tube and an abruptly flared flat limb, as in *Primula* or *Phlox*

samara (L., *samara,* the fruit of the elm)- an indehiscent, single-seeded, dry fruit with a prominent wing, as in the fruits of the elm, ash, or tree-of-heaven; a prominently winged achene

saprophyte (Gk., rotten + a plant)- a plant which derives its nourishment from dead organic material in the soil

scabrous (L., *scaber,* rough)- rough to the touch because of bristly hairs

scale (ME, a husk)- a general term applied to any small, dry, vestigial leaf or bract

scandent (L., *scandens,* climbing)- climbing; in a more restricted sense, climbing without the aid of tendrils

scape (L., *scapus,* a stem)- a peduncle which lacks any well-developed foliage leaves, although bracts may be present; **scapose,** having a scape

scarious (L., *scaria,* a thorny shrub)- thin, dry membranous, and non - green; applied particularly to the margins of leaves and bracts

schizocarp (Gk., I split + a fruit)- a dry indehiscent fruit derived from a syncarpous gynoecium which splits at maturity into 1-seeded closed segments **(mericarps);** in its most restricted usage the term is applied to the fruit of the Umbelliferae; considered indehiscent even though it splits open because segments remain closed and do not shed seeds

scurfy (ME, *scurf,* ?)- covered with minute scales

seed (L., *serere,* to sow or plant)- the fertilized and ripened ovule

sepal (Gk., a covering)- one of the component parts of the calyx

sepaloid- resembling a sepal, as in some petals

septate- divided by a partition, as in an ovary or fruit

septicidal capsule (L., *septicidalis,* an enclosure + I cut)- a capsule which dehisces by means of openings along or within the septations; see also **loculicidal capsule**

septum (L., *septum,* a hedge or enclosure)- a partition, as within an ovary

sericeous (L., *sericeus,* silky)- silky, the hairs fine, long, and appressed

serrate (L., *serra,* a saw)- having sharp forward pointed teeth, as on a leaf margin

serrulate- minutely serrate

sessile (L., *sessilis,* sitting)- without a stalk; seated directly on the supporting structure or substrate

seta (L., *seta,* a bristle)- a bristle; **setaceous,** bristle-like

setose- covered with bristles

sheath (ME, *schethe,* to cut or split)- an elongate tubular structure which surrounds an organ or plant part

shrub (ME, *schrubbe,* brushwood)- a woody plant with few to several trunks from the base

silicle (L., *silicula,* a little husk or pod)- the fruit of certain Cruciferae; a short fruit (typically not more than 2-3 times longer than wide) which dehisces in such a way that a central papery partition remains after the fruit walls have fallen; see also the next definition

silique (L., *siliqua,* a pod)- the elongate, bicarpellate, longitudinally dihiscent fruit of the Cruciferae; the walls of the fruit fall away revealing a persistent internal partition (the **replum)**

simple leaf- a leaf which is not compounded into discrete leaflets

simple pistil- a pistil consisting of a single stigma, style, and placenta, as in the unicarpellate gynoecium or each of the carpels in an apocarpous gynoecium

sinus (L., *sinus*, a curve or fold)- a recess between two lobes, as in the space defined by two lobes of a leaf

sorus (Gk., a heap)- a group of sporangia which are clustered as on the undersurface of a fern leaf

spadix (Gk., a palm branch with fruits)- a spike of small, crowded, often unisexual flowers on a fleshy axis; the inflorescence typical of the Araceae, but not limited to it

spathaceous- spathe-like

spathe (Gk., a spatula)- a large bract or leaf which subtends and often partially surrounds an inflorescence; it is often found in association with the spadix

spatulate (L., *spatula*, a spoon)- shaped like a druggist's spatula; ± spoon-shaped

species (L., *species*, a kind or sort)- a kind of plant or animal, its distinctness seen in morphological, anatomical, cytological, and chemical discontinuities presumably brought about by reproductive isolation; thought by some to be entities with biological reality and by others to be convenient concepts which exist only in the mind of the taxonomist

species name- the binomial consisting of the genus and the specific epithet

specific epithet- the second element of a scientific name; often loosely referred to as the species name

spicate- spike-like

spike- (L., *spica*, an ear of corn)- an elongate, unbranched, indeterminate inflorescence whose flowers are sessile

spikelet (L., *spicula*, dim. of *spica*)- a spicate arrangement of reduced flowers and associated bracts inserted alternately along an unbranched axis (the **rachilla**); a small spike; the basic unit of the inflorescence in the grasses and sedges

spine (L., *spina*, a thorn)- a sharp - pointed leaf or portion of a leaf, as in the stipular spines of *Acacia;* used by some authors instead of **thorn,** but here distinguished from it by its leaf origin

spinescent- ending in a spine or sharp point

spinose-with a spine at the tip

sporangiophore (Gk., seed + vessel + I carry)- the stalk of a sporangium

sporangium (Gk., seed + vessel)- a spore-producing structure

spore (Gk., seed)- a reproductive cell which can be released and is capable of developing into a new individual

sporocarp (Gk., seed + fruit)- the bean-like receptacle which contains the sporangia of heterosporous ferns, such as *Salvinia* and *Marsilea*

sporophyll (Gk., seed + leaf)- a spore - bearing leaf, as in those leaves in lower vascular plants which bear sporangia, or in a more heavily modified form, the stamen and carpel of the flowering plants

spreading- oriented outward and ± diverging from the point of origin

spur (L., *spernere*, to push away)- a hollow slender or sac-like extension of some part of the flower, as in a sepal of *Delphinium* or a petal of *Viola*

stamen (L., *stamen*, a filament)- the basic unit of the androecium; the pollen-producing organ of the flower; the male part of the flower; generally thought to be homologous with a microsporophyll

staminate-a flower which contains a functional androecium, but which lacks a well-developed functional gynoecium; a male flower

staminode (= **staminodium**)- a sterile or abortive stamen or a structure interpreted as such because of its position within the flower; may closely resemble a stamen or be heavily modified and petaloid

standard (Fr., *estendard*, a standing place)- the upper (fifth or posterior) petal in a papilionaceous corolla, often bilobed and thus appearing as two petals; the upper petal (enclosed in bud by two lateral ones) in a caesalpiniaceous corolla; any of the three petals of an *Iris* flower

stellate (L., *stellatus*, starry)- radiating like the points of a star, as in some hairs with radiating side branches

stem (Me, *stemn*, tree trunk)- the leaf, flower, and fruit bearing axis of a plant; principally aerial, but also subterranean in the form of rhizomes, corms, etc.

sterile (L., *sterilis*, barren)- without functional stamens or carpels

stigma (Gk., a point)- the point or region of the gynoecium which is receptive to pollen; usually apical and in the form of a point, disc, lobe or line; **stigmatic**, of or pertaining to the stigma

stipe (L., *stipes*, a trunk)- the stalk of a gynoecium within a flower, as in the Capparaceae; the petiole of a fern leaf

stipule (L., *stipula*, stubble)- an appendage, usually seen as paired structures, inserted at the base of the petiole; please note that the stipules are part of a leaf and not bracts subtending it

stolon (L., *stolo*, a shoot)- a horizontal stem, usually above the surface of the ground, which roots at the nodes and produces new plants, particularly at its tip; also referred to as a **runner** or **sucker**

stone (ME, *ston*, to become thick)- the bony endocarp and seed within the drupe of plums, peaches, and related genera of the Rosaceae; see also **pyrene**

stone fruit- a fruit which contains a stone

striate (L., *stria*, a furrow)- marked with fine longitudinal parallel lines, groves or ridges

strict (L., *strictus*, drawn together)- close, narrow, upright, very straight

strigose (L., *strigosus*, meager)- covered with sharp, straight, stiff, appressed hairs

strobilus (L., *strobilus*, anything twisted)- a cone; a series of sporophylls on a common axis

style (Gk., a column)- the elongate neck of the gynoecium; the portion between the ovary and the stigma

stylopodium (Gk., column + foot)- the enlargement at the base of the styles in the Umbelliferae

subglobose (L., *subglobosus*, nearly globose)- almost spherical

submersed (L., *submersus*, plunged under)- growing under water; submerged

subtend (L., *subtendo*, I stretch underneath)- to be situated below, as in a bract inserted beneath a flower

subulate (L., *subula*, a small weapon)- the shape of a leaf or bract when it tapers from the base to the apex as in an awl, a carpenter's tool for piercing leather or wood; also referred to as **awl-shaped**

succulent (L., *succulentus*, sappy)- juicy, fleshy, thickened; a plant having fleshy leaves and/or stems

suffrutescent (L., *suffrutescens*, somewhat + a shrub)- somewhat shrubby, as in a plant which is woody toward its base, but herbaceous above; also referred to as **suffruticose**

sulcate (L., *sulcatus,* furrowed)- grooved or furrowed

superior (L., *superior,* higher)- above or on top of, as in the superior ovary which is inserted above the points of insertion of the perianth and androecium

suture (L., *sutura,* a seam)- a line or groove which marks a place of union or separation, as in the capsule types which open along sutures

syconium (Gk., fruit of the fig tree)- the false fruit of the fig; a vase - like structure with an opening at its apex and whose interior wall is lined with tiny flowers

sympetalous (Gk., with + a flower leaf)- with united petals; also referred to as **gamopetalous**

synangium (Gk., growing together + a vessel)- the structure which results from fusion of sori

syncarpous (Gk., with + a fruit)- said of a gynoecium with two or more fused carpels

syngenesious (Gk., with + connate)- the condition of the androecium when the stamens are united by their anthers, as in the Compositae

synsepalous (Gk., with + a covering)- with united sepals; also referred to as **gamosepalous**

tailed (AS, *taegel,* ?)- any slender basal prolongation of a structure, particularly an anther

tap root- a primary descending root, larger than the others of the root system

taxon (Gk., arrangement)- a taxonomic group of any rank; plural, **taxa**

taxonomic hierarchy (L., *hierarcha,* keeper of sacred things)- the ordered list of categories into which plants are classified

tendril (Fr., *tendre,* to stretch out)- a thread - like structure, of stem or leaf origin, by which a plant supports itself

tepal (L., *tepalum,* anagram of *petalum*)- a component part of a perianth which is not differentiated into a calyx and corolla; used when it is not possible to determine easily whether a structure is a sepal or a petal

terete (L., *teres,* rounded)- circular in cross - section

ternate (L., *terni,* in 3's)- occurring in 3's, as in leaflets of a compound leaf

tetradynamous (Gk., four + power)- having four long stamens and two short ones, as in the androecium of the Cruciferae

thallus (Gk., a sprout)- a plant body which is not differentiated into roots, stems, and leaves, as in plants of the Lemnaceae

theca (Gk., a case)- a pollen sac or half - anther

thorn (ME, to be stiff)- a sharp - pointed stem; may be simple or branched; see also **spine**

throat (ME, *throte*)- the opening into a sympetalous corolla

thyrse (L., the staff of Bacchus)- a condensed panicle-like inflorescence, in which the main axis is indeterminate and the lateral branches are determinate

tomentose (L., *tomentum,* cushioning)- covered by a dense, soft, woolly hariness

tomentulose- minutely tomentose

torus (L., *torus,* a bed)- the receptacle of a flower

trailing- creeping or sprawling plants, often

rooting at the node; also referred to as **repent**

tree (AS, *treow*, a tree)- a perennial woody plant with a single evident trunk from its base

trichome (Gk., a growth of hair)- any hair-like outgrowth of the epidermis

trifoliate (L., *trifoliatus*, three-leaved)- a plant with three leaves, as in *Trillium*

trifoliolate (L., *trifoliolatus*, three leaflets)- having three leaflets

tripinnate- three times pinnately compound

truncate (L., *truncatus*, shortened)- a straight base or apex which appears to have been cut off

tuber (L., *tuber*, a tumor)- a swollen underground stem tip, as in the Irish potato

tuberculate. see **verrucose**

tuberous root- a tap root which is particularly large and fleshy, as in the table beet

turion (L., *turio*, a shoot)- a swollen scaly offshoot of a rhizome

twining- coiling around plants or objects as a means of support

true fruit- those fruits derived from the syncarpous or unicarpellate gynoecium of a single flower

translator (L., *translator*, a transferrer)- the arm which connects the pollinia to the corpusculum in the milkweed flower; also referred to as a **connective**

umbel (L., *umbella*, a sunshade)- a flat-topped or rounded indeterminate inflorescence in which all of the flowers are borne on pedicels of ± equal length and which arise from a common point

uncinate (L., *uncinatus*, hooked)- hooked

undulate (L., *undulatus*, wavy)- wavy perpendicular to the plane of the blade; wavy up and down, rather than in and out, as in **sinuate**

unilocular (L., *uni* + *loculus*, one + a small compartment)- having a single locule or chamber

uniseriate (L., *uniseriatus*, one + a row)- a perianth of a single series

unisexual (L., *unisexualis*, one sex)- a flower or a plant which bears either stamens or carpels, but not both

urceolate (L., *urceolatus*, relating to pitchers)- urn - shaped; said of a corolla which is wide at the bottom and contracted at the mouth

utricle (L., *utriculus*, a small skin or husk)- an indehiscent, one-seeded, bladdery fruit

valid publication - the publication of a new scientific name, in order to be valid, must include an original Latin description of the new taxon or a reference to a previously published one

valvate (L., *valvatus*, with folding doors)- meeting without overlapping

valve (L., *valva*, the leaf of a door)- the pieces into which a capsule, silicle or silique separate after dehiscence has occurred

velutinous (L., *velutinus*, velvety)- covered with a velvety hairiness

venation (L., *vena*, a vein)- the arrangement of veins

ventricose (L., *ventricosus*, relating to the belly)- a swelling which occurs on one side, as in the corolla of certain mints and scrophs

versatile (L., *versatilis*, movable)- moving about free on its support, as in an anther on its filament

verticil (L., *verticillus,* the whirl of a spindle)- a whorl of similar parts about an axis, used particularly to describe a whorl of flowers at a node, as in the mints; **verticillate,** whorled

vestiture (= **vesture** and **investiture**) (L., *vestire,* to clothe)- any surface feature (hairs, scales, etc.) on a plant part

villous (L., *villus,* a shaggy hair)- covered with shaggy, soft, but not matted hairs

viscid (L., *viscidus,* clammy)- sticky

viscidium- the sticky disc to which the pollinia are attached in the Orchidaceae

vine (L., *vinea,* a vine)- herbaceous plants with elongate, flexible, non-self-supporting stems; see also **liana**

vittae (L., *vitta,* a fillet)- the oil tubes in the pericarp of most members of the Umbelliferae

wanting (ME, *wanten,* to be lacking)- lacking

whorl (ME, *whorwyl,* variation of whirl)- a circular arrangement involving three or more leaves or flowers at a node

wing (ME, *winge;* AS, *fethra,* wing)- a thin, dry, membranous expansion of an organ or structure, as on the surface of a samara, silique or schizocarp; a lateral petal in certain legume flowers

xerophyte (Gk., dry + a plant)- a plant of arid habitats

zygomorphic (Gk., a yoke + shaped)- having a symmetry such that only one plane passed through the flower will yield two mirror images; also referred to as **bilaterally symmetrical**

Generic names have not been included in this index. Family names appearing in italics are synonyms or are rejected for other taxonomic reasons.

Index